9th International Conference on Magnetic and Superconducting Materials 2015 (MSM15)

Journal of Physics: Conference Series Volume 667

Antalya, Turkey
30 April – 3 May 2015

Editors:

M. Eyyuphan Yakinci
Bilal Tanatar

ISBN: 978-1-5108-1818-7
ISSN: 1742-6588

Printed from e-media with permission by:

Curran Associates, Inc.
57 Morehouse Lane
Red Hook, NY 12571

Some format issues inherent in the e-media version may also appear in this print version.

Printed by Curran Associates, Inc. (2016)

For permission requests, please contact the Institute of Physics
at the address below.

Institute of Physics
Dirac House, Temple Back
Bristol BS1 6BE UK

Phone: 44 1 17 929 7481
Fax: 44 1 17 920 0979

techtracking@iop.org

Additional copies of this publication are available from:

Curran Associates, Inc.
57 Morehouse Lane
Red Hook, NY 12571 USA
Phone: 845-758-0400
Fax: 845-758-2634
Email: curran@proceedings.com
Web: www.proceedings.com

Table of contents

Volume 667

9th International Conference on Magnetic and Superconducting Materials (MSM15)
30 April to 3 May 2015, Antalya, Turkey

Accepted papers received: 18 November 2015
Published online: 5 January 2016

Papers

The *Magnetic and Superconducting Materials (MSM) Conference*, which is held biannually in many countries, started in 1999 as a scientific conference on magnetism and superconductivity and their application in materials. Today *MSM* conferences aim to bring together regional and international research groups and technologists of related companies to discuss new materials, their results and current problems in the areas of superconductivity, magnetism, novel materials and also in the general field of Low Temperature Physics. The main goal of the *MSM* conference series is to increase collaboration within the region and the third world countries, and with the international community among the scientists from the developed countries. It is only through the sharing of experience with scientists around the world that one could fully benefit from natural and human resources. During the *MSM15* conference we have received more than 250 abstracts from 22 different countries, plenary, invited and contributed talks were presented and many scientific subjects were widely discussed. The contributions in this volume have been reviewed by the eminent international scientists and represent some of the invited and contributed talks presented during *MSM15* conference. We would like to thank to all of the participants attending the conference and also international scientific committee for their contribution to a high level conference and its overall success. We also would like to thank to our sponsors İnönü University, Gazi University, Çukurova University, Abant İzzet Baysal University and Bülent Ecevit University and more than eight scientific companies during the conference.

M. Eyyuphan Yakinci, Bilal Tanatar

Editors

Peer review statement

All papers published in this volume of *Journal of Physics: Conference Series* have been peer reviewed through processes administered by the proceedings Editors. Reviews were conducted by expert referees to the professional and scientific standards expected of a proceedings journal published by IOP Publishing.

Thermal Conductivity and Thermoelectric Power of Yb-Substituted Bi-2212 Superconductor

H Gündoğmuş [1,2*], B Özçelik[2], A Sotelo[3] and M A Madre[3]

[1] Faculty of Engineering, Hakkari University, 30000 Hakkari, Turkey
[2] Department of Physics, Faculty of Sciences and Letters, Çukurova University. 01330 Adana, Turkey
[3] ICMA (CSIC-Universidad de Zaragoza) María de Luna, 3. 50018 Zaragoza, Spain

E-mail: hakangundogmus@hu.edu.tr

Abstract. In the present study, $Bi_2Sr_2CaCu_{2-x}Yb_xO_{8+y}$ where x = 0.0, 0.05, 0.1and 0.25, superconductor materials were textured using the Laser Floating Zone technique. After annealing the as-grown materials, they were characterized by thermal conductivity (*κ-T*), thermoelectric power (*S-T*), and electrical resistivity (*R-T*) measurements performed from room temperature down to 20 K. The thermoelectric power is positive in all cases and is found to increase with decreasing temperature, reaching the maximum value (peak) around the samples *Tc* values and dropping rapidly to zero below *Tc*. Electrical resistivity measurements show a slight decrease of T_C when Yb substitution is raised. In spite of this effect, T_C values are higher than 90 K in all cases. On the other hand, Yb substitution also decreases thermal conductivity, compared with the undoped samples.

1. Introduction

Since Bi-based superconductors $Bi_2Sr_2Ca_{n-1}Cu_nO_x$ (n = 1, 2, 3) were discovered by Maeda *et al.* [1] many substitutions have been made in BSCCO system in order to improve their physical and magnetic properties [2-14]. High-T_C Bi-based superconductor family is characterized by a high superconducting transition temperature T_C ~ 90 and 110 K for the phases with n = 2, and 3, respectively, and a high upper critical magnetic field, B_{c2}, which can reach 150 T [15]. The weak coupling between BiO-BiO layers in the BSCCO system enables substitution of rare-earth ions for Ca and/or Bi and makes the system interesting from the point of studying its structural, electrical, thermal and magnetic properties. It is well-known that the BSCCO system superconducting properties are very sensitive to the hole-carrier concentration which, in turn, depends on the substitution levels [16, 17].

In general, the studies of transport properties, such as electrical resistivity, thermoelectric power (S) and thermal conductivity (κ), are important for exploring the conduction mechanism and the nature of charge carriers. The thermoelectric properties of BSCCO have been investigated by many research groups [18-23]. It was reported that S values have been found to be positive, negative or both, depending on the substituting element and proportion. The presence of both positive and negative values of the thermoelectric power indicates the effect of two types of charge carriers, i.e. holes and electrons, respectively. In addition, the material preparation techniques for the BSCCO systems play a crucial role. Laser Floating Zone Melting (LFZ) technique is one of the most reliable methods to texture these bulk superconductor samples [24]. This method provides well aligned Bi-2212 grains, with their c-axis nearly perpendicular to the growth direction, and very small amount of secondary phases. In previous works, the structural, physical, magneto resistance and superconducting properties of Yb-substituted BSCCO samples have been studied [6, 25]. In the present study, the electrical resistivity, thermoelectric power and thermal conductivity of LFZ textured $Bi_2Sr_2CaCu_{2-x}Yb_xO_{8+y}$ (x = 0.0, 0.05, 0.1, and 0.25) superconductors are presented. Moreover, with these data, the figure-of-merit, ZT ($=S^2T/\rho\kappa$) has been calculated to determine the thermoelectric performance of these samples.

2.Experimental Details

$Bi_2Sr_2CaCu_{2-x}Yb_xO_{8+y}$ samples, with x = 0.0, 0.05, 0.1, and 0.25, have been prepared by using a polymer matrix route described in detail in previous works [6]. In order to identify the present phases, powder X-ray diffraction patterns of the materials were recorded at room temperature using a Rigaku D/max-B powder diffract meter system working with CuK_α radiation and a constant scan rate between 2θ= 3-60°. Resistivity and thermal characterization measurements were carried out in a Quantum Design PPMS system. The $Bi_2Sr_2CaCu_{2-x}Yb_xO_{8+y}$ samples, with x = 0.0, 0.05, 0.1, and 0.25 will be hereafter named as A, B, C and D, respectively.

3.Results And Discussion

Normalized XRD diffraction patterns of A, B, C and D samples are displayed in Fig. 1. All samples mainly consist of Bi-2212 phase which indicates that the annealing process has been adequate. However, some impurity phases like $CaCuO_2$ were also detected. Unit cell parameters and particle sizes of samples are given table 1. As seen in Table 1, with increasing Yb content, a non-monotonic decrease of the c-parameter with simultaneous increase of the a-one was observed. Particle size calculations have been performed using Debye Scherer Formula [26].

$$L_{hkl}= 0,9\ \lambda\ /\ \beta\cos\theta \qquad (1)$$

where λ is the wavelength of the source, β is the full width at half maximum and θ is the angle. An indication of grain size can be roughly obtained from the peak width, if the diffraction peak is narrow, the crystal size is large. In Table 1, by using Eq.1, it can be seen that the particle sizes decrease with increasing Yb content.

Table 1. Unit cell parameters, particle size and hole-carrier concentration for the samples

Sample	a=b (Å)	c(Å)	L_{hkl} (Å)	*p*
A	3.834	30.950	371.1391	0.144
B	3.832	30.939	364.5567	0.138
C	3.832	30.900	332.2064	0.135
D	3.832	30.850	319.1931	0.135

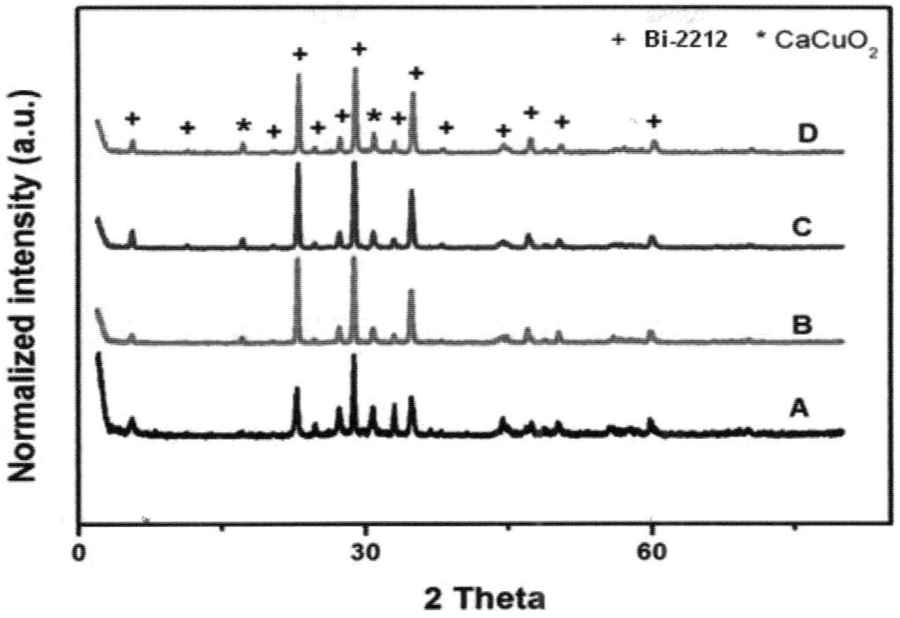

Figure 1. X-ray diffraction patterns of $Bi_2Sr_2CaCu_{2-x}Yb_xO_y$ superconducting samples

Fig. 2 shows the temperature dependence of the resistivity. The ρ curve shows a linear dependence with respect to the temperature between room and transition temperatures with metallic behavior. On the other hand, it can be clearly seen that T_c is approximately the same for all the samples. It is valuable to say that the increase of the resistivity in metallic region is due to the decrease of the grain alignment and the modification of grain size together with the appearance of secondary phases, as can be seen at XRD patterns. T_c^{onset} and T_c^{offset} temperatures are found as 100, 98.6, 97.7, and 96 K, and 95, 93.2, 92.1, and 90 K, for A, B, C and D samples, respectively. As a consequence, it is clear that Yb decreases T_C values but in a small extent. The temperature dependence of S for all the samples is given in Fig.3. All the samples exhibit positive S values indicating hole like conduction.

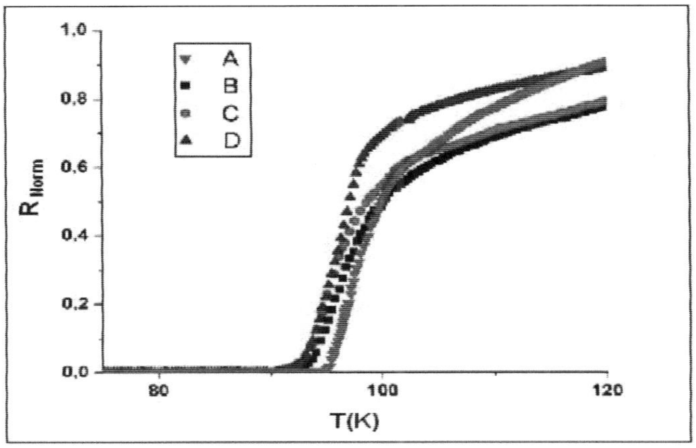

Figure 2. Temperature dependence of normalized resistivity for all the samples

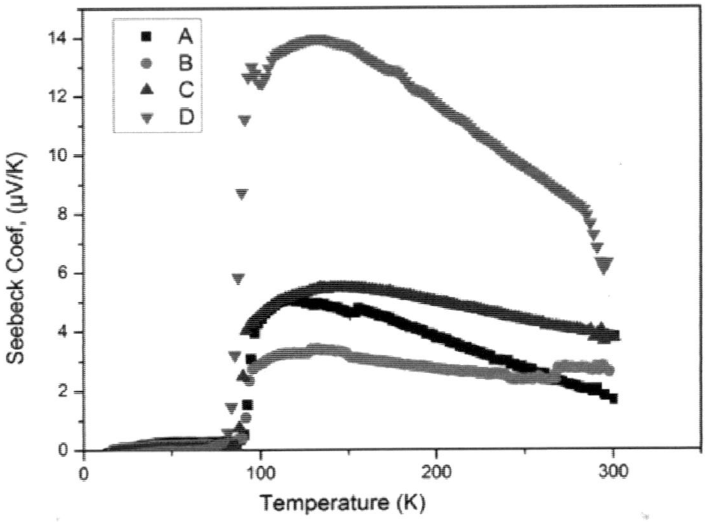

Figure 3. Temperature dependence of thermoelectric power for all the samples

This is similar to those observed for the substituted BSCCO system reported in other works [23, 27]. The positive *S*-values obtained in these samples are caused by the contribution of Bi-2212 phase. Furthermore, *S* increases by decreasing the temperature, and attain maximum values at around their respective transition temperature values, before dropping to zero at lower ones, as expected. Moreover, in spite of the anomalous behavior of B sample, as a general trend it can be deduced that the magnitude of *S* increases with Yb-content.

In Fig. 4, the temperature dependence of thermal conductivity for all the samples is exhibited. As can be observed, for $T>Tc$ (metallic state), κ first linearly increases from room temperature, reaching a maximum and then decreases with decreasing temperature. Around Tc, first a shallow minimum was observed following by a small raise. Then with decreasing T, almost a linear decrease towards zero was found.

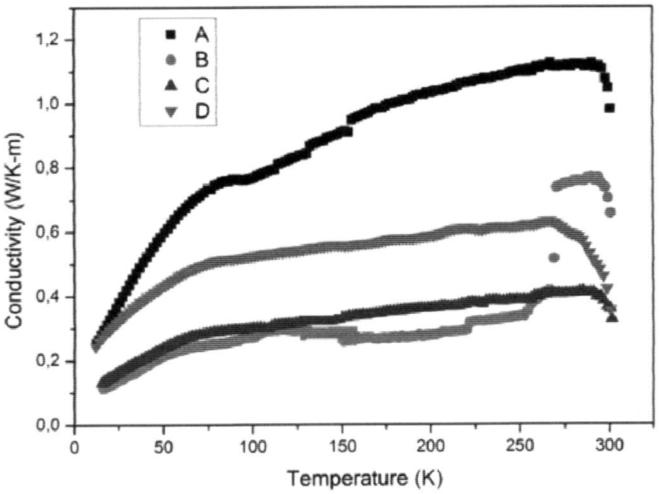

Figure 4. Temperature dependence of the thermal conductivity for all the samples

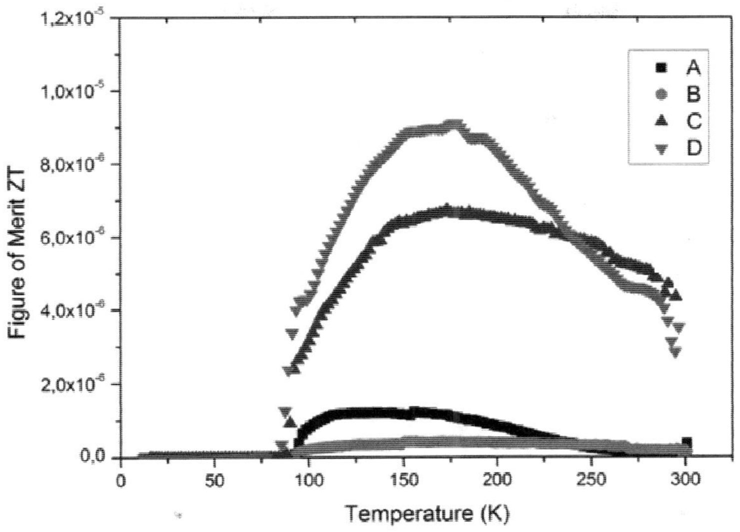

Figure 5. Figure of Merit for the samples

This type of behavior has previously been reported in the H*Tc* systems [22, 23, 28, 29]. The origin of the minimum can be attributed to the superconducting fluctuations [30, 31]. These results reveal that the magnitude of the thermal conductivity, κ, is influenced by the Yb- concentration in the system. Namely, with increasing Yb concentration the magnitude of κ is decreased. The dimensionless figure of merit, ZT, is defined as:

$$ZT = \alpha^2 T/\rho\kappa \qquad (2)$$

where T, α, ρ, κ are temperature, thermoelectric power, electrical resistivity and thermal conductivity, respectively. According to Eq.2, ZT is maximized when the thermoelectric power increases and the electrical resistivity and thermal conductivity decrease. As can be seen in Fig. 5, ZT is maximized at the temperature where thermoelectric power is maximized. Moreover, it becomes zero below the transition temperature of each compound. Furthermore, it is clear that the general trend is increasing ZT with Yb content.

4.Conclusions

Textured $Bi_2Sr_2CaCu_{2-x}Yb_xO_y$ superconducting materials with x= 0.0, 0.05, 0.1, and 0.25 were successfully prepared by the LFZ technique. Electrical resistivity measurements show that Yb substitution slightly decreases T_c. Nevertheless, the minimum measured value is still higher than 90 K. Thermoelectric power is increasing with Yb content while thermal conductivity is decreased. All these factors lead to an important raise of ZT for the Yb-doped materials which is maximum for the 0.25Yb-doped samples. Moreover, the improvement of ZT is about 5 times the measured in undoped samples.

Acknowledgements
This work is supported by Research Fund of Çukurova University, Adana, Turkey, under grant contracts no: FEF 2011D21.

References

[1] Maeda H *et al.* 1988 *Jpn. J. Appl. Phys.* **27** L 209
[2] Özçelik B *et al. Mater. Sci: Mater Electron.* **25** 2456
[3] Özçelik B *et al.* 2014 *J. Low Temp. Phys.* **174** 136
[4] Türk N *et al.* 2014 *J Supercond. Nov. Magn.* **27** 711
[5] Özaslan A *et al.* 2014 *J. Supercond. Nov. Magn.* **27** 53
[6] Gündogmus H *et al. 2013 J. Supercond. Nov. Magn*, **26** 111
[7] Ozcelik B *et al.* 2014 J. *Mater. Sci: Mater Electron.* **25** 4476
[8] Özçelik B *et al.* 2013 *J. Supercond. Nov. Magn.* **26** 873
[9] Yazıcı D *et al.* 2011 *J. Low Temp. Phys.* **163** 370
[10] Özkurt B *et al.* 2012 J. *Supercond. Nov. Mag.* **25** 799
[11] Ozcelik B *et al.* 2015 *J. Supercond. Nov. Magn.* **28** 553
[12] Ozcelik B *et al.* 2015 *J. Mater. Sci: Mater Electron.* **26** 441
[13] Ozcelik B *et al.* 2015 *J. Mater. Sci: Mater Electron DOI* **10.**1007/s10854-015-2765-1
[14] Gursul M *et al.* 2015 *J. Supercond. Nov. Magn.* DOI **10.**1007/s10948-015-2977-x
[15] Blatter G *et al.* 1994 *Rev. Mod.Phys.* **6 1**125
[16] Torrance J B *et al.* 1998 *Phys.Rev. Lett.* **61** 1127
[17] Tokura Y *et al.* 1998 *Phys. Rev.* B **38** 7156
[18] Chatterjee S *et al.* 1996 *Phys. Rev.* B **53** 5942
[19] Yakinci M E *et al.* 2007 *Physica C* ***460*** 1386
[20] Bhattacharya S 1998 *J. Mater. Sci. Lett.* **17** 1575
[21] Aksan M A *et al.* 2004 *J. Alloy Comp.* ***385*** 33
[22] Ozkurt B *et al.* 2007 *Physica C.* **467** 112
[23] Ozkurt B *et al.* 2007 *Low Temp.Phys.* **147** No:1/2 April 31
[24] De la Fuente G F *et al.* 1995 *Adv. Mater.* **7** 853
[25] Yakinci M E 1997 *J. of Phy. Con. Matt.* **9** 1105
[26] Cullity B D 1978 *Element of X-ray Differaction, Addition-Wesley, Reading, MA*
[27] Kishore K N *et al.* 1995 *Physica C.* **252** 49
[28] Chanda B *et al* 1994 *Solid State Commun.* **89** 353
[29] Aksan M A *et al.* 1999 *J. Low Temp. Phys.* **117** 957
[30] Cohn J L *et al.* 1992 *Phys. Rev.* **B 45** 13144
[31] Castellazi S *et al.* 1997 *Physica C.* **273** 314

Journal of Physics: Conference Series **667** (2016) 012002 doi:10.1088/1742-6596/667/1/012002

The physical and magnetic properties of FeSe-11 superconductor

C Işık, N Kantarcı Güler, M Gürsul, A Ekicibil and B Özçelik

Department of Physics, Faculty of Sciences and Letters, Çukurova University, Adana, Turkey

E-mail: ozcelik@cu.edu.tr

Abstract. In this study, polycrystalline of $FeSe_{0.98}$ and $FeSe_{1.2}$ samples were synthesized by a solid-state reaction method to investigate the microstructural and magnetic properties in these systems. Different heating cycles were applied and the best synthesis cycle has been established by optimization. Samples were characterized regarding structure, composition and superconducting properties. The microstructural investigations showed that structural formation of the samples was not strictly depending on the heating cycles. Transition temperature is obtained as 10 K from Zero Field Cooled magnetization measurement.

1. Introduction

During the past seven years, a new type of Fe-based superconductor, tetragonal FeSe with a superconducting transition temperature T_c of 8K, was reported [1]. As observed in the cuprates, the Fe-based superconductors exhibit interplay between magnetism and superconductivity suggesting the possible occurrence of unconventional superconducting states. Other general properties of these superconductors are the layered structure and the low carrier density. Among the Fe-based superconductors, FeSe has the simplest structure with layers in which Fe cations are tetrahedrally coordinated by Se anions [2].

The report describing the superconductivity in the tetragonal FeSe suggests that superconductivity is observed in samples prepared with significant selenium deficiency. The stoichiometry of the superconducting phase is reported to be between $FeSe_{0.82}$ and $FeSe_{0.88}$ [3]. The first powder neutron diffraction experiment also argue that selenium vacancies are present, proposing a stoichiometry of $FeSe_{0.92}$ in that particular sample [4]. Theoretical treatments of this phase have either assumed the stoichiometry to be FeSe [1,5], without accounting for the critical differences in electron count of 0.16-0.3 Se/Fe that would be present for selenium-deficient materials, or have specifically address the importance of the vacancies in giving rise to the superconductivity [6,7]. Many measurements have been reported on samples that have been prepared with the assumption that iron selenide is deficient in Se [8-12]. Many preparation techniques can be used for the synthesis of FeSe samples. These techniques are solid-state reactions, high pressure synthesis, flux growth and Bridgman methods [13]. In generally the first and second methods are used to prepare polycrystalline samples while the latters are used for the growth of FeSe single crystals.

As we have mentioned above, the superconductivity in β-FeSe is very sensitive to composition and disorder, since the nonstoichiometry and intrinsically Se deficient nature plays crucial role in the matrix. A reproducible sample fabrication procedure and prevention from the oxidation are needed

Journal of Physics: Conference Series **667** (2016) 012002 doi:10.1088/1742-6596/667/1/012002

during the preparation process. Therefore, in this study, polycrystalline of $FeSe_{0.98}$ and $FeSe_{1.2}$ samples were synthesized by solid-state reaction method to study the microstructural, superconducting properties in these systems. Different heating cycles were applied and the best synthesis cycle has been established by optimization. We emphasize here that almost single phase of FeSe can only be obtained at near a perfect 1:1 ratio of Fe to Se.

2. Experimental

Polycrystalline $FeSe_{1.2}$and $FeSe_{0.98}$ samples were prepared by solid-state reaction method in sealed quartz tubes using high purity powders of the iron (99.999 % Alfa Aesar) and selenium (99.999 % Alfa Aesar) with appropriate stoichiometries of Fe/Se:1/1.2 and 1:0.98, respectively. Initially well homogenized raw powders were placed into quartz tube and sealed in the glove box. For synthesis of $FeSe_{1.2}$a quartz tube containing the elements in the corresponding ratio was heated up to 700^0C with 3 different durations, namely 12, 24 and 36 hours. For each heating cycles, constant heating and cooling ramps of 2^0C/min were used. In each step, intermediate grinding were applied in glove box under Ar atmosphere and then sealed into an evacuated quartz tubes. For the last heating stage samples were pulverized again and pressed isostatically in to a pellet of 8 mm diameter by applying an 80 kN pressure in the glove box.

For synthesis of $FeSe_{0.98}$, the mixture of the elements in the corresponding ratio was heated up to 780 ^{0}C with a heating ramp of 1^0C/min where it was kept for 48 h. This high temperature step is required for scrubbing oxygen from the system and obtaining homogeneous product. The furnace was cooled down to room temperature with a cooling ramp of 5 ^{0}C/min. Then the powder was ground and pressed again isostatically in to a pellet of 8 mm diameter by applying an 80 kN pressure in the glove box and put again in the furnace. The sample was heated up to 430 ^{0}C with a heating ramp of 20 ^{0}C/h where it was kept 96 h. This step is needed to avoid the low temperature decomposition of superconducting phase. Finally, the sample was taken out after the furnace cooling to room temperature.

After the heat treatment cycles, structural properties of the materials obtained were examined by X-Ray Powder diffraction. A Rigaku RadBD MaxII difractometer and Jade 5+ refinement program was used to analyze structure. Structural parameters were calculated by using Rietveld refinement based on the Jade 6+ crystal refinement program. The microstructures of samples were investigated by using Scanning Electron Microscopy (SEM; Leo EVO 40VPX) and Energy Dispersive X-ray spectroscopy (EDX; Bruker 125 eV). The superconducting properties of the FeSe polycrystalline samples were carried out using Physical Properties Measurement Systems, (Quantum Design; PPMS-9T). Up to 9 Tesla were applied during the magnetization measurements.

3. Results and discussions

The XRD patterns of both samples are given in Fig. 1. The broad halo around $15^o \leq 2\theta \leq 30^o$ in the XRD pattern of the $FeSe_{0.98}$ sample is caused by a sample holder with a plastic dome. We realized significant difference between the patterns of the samples. The peak intensities of $FeSe_{1.2}$ are sharper than $FeSe_{0.98}$ indicating much higher crystallinity. While $FeSe_{1.2}$ sample has no different phase, $FeSe_{0.98}$ sample contains an impurity arising un-reacted Fe ions remains in the compound during the formation of FeSe. This may be related to the difference of melting temperature between the Fe (1521 ^{0}C) and Se (221 ^{0}C) [14]. Even their peak intensities are different; they have exhibited same crystallization behavior. The crystal symmetry was found to be tetragonal with the space group of *P4/nmm* and crystal parameters were calculated to be a= 3.77361(6) and c=5.52582(5). These results are in agreement with the results obtained by various research groups [15-17].

Fig. 2 shows the surface micrographs of samples. Randomly oriented granular formation with an average grain size of 3-4 μm is observed for two samples. This is general characteristic of Fe-based superconductors. $FeSe_{0.98}$ and $FeSe_{1.2}$ samples have similar microstructural formation, indicating that there is no significant effect of preparation process and thermal treatment on the morphological evolution.

Journal of Physics: Conference Series **667** (2016) 012002 doi:10.1088/1742-6596/667/1/012002

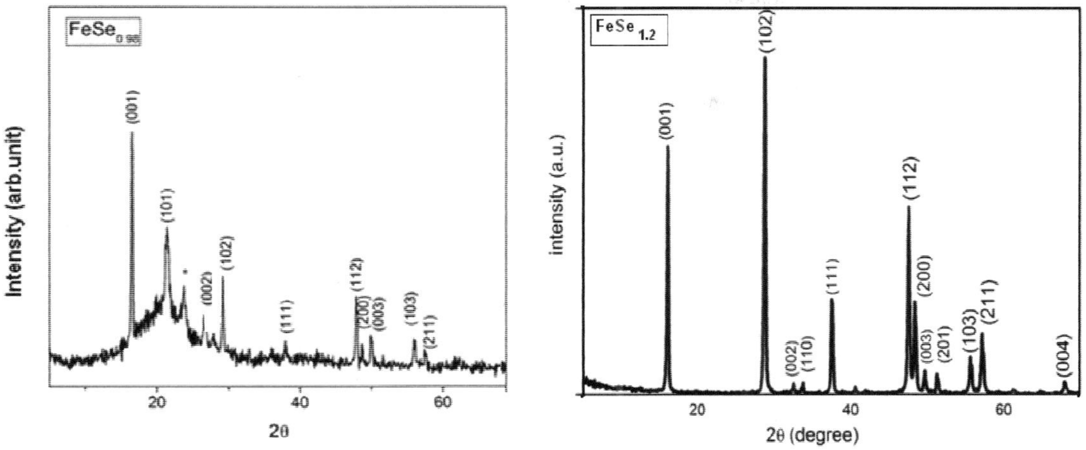

Figure 1. XRD results of the samples

Large scale EDX dot mapping result of $FeSe_{0.98}$sample is given in Fig. 3. As can be seen in Fig. 3, distributions of the FeSe phase on the surface of the sample are almost homogenous. In addition, any deficient distribution of Fe and Se atoms and/or non superconducting phases on the surface of the sample has not been observed. According to EDX analysis, atomic concentration of the samples was found very close to their nominal composition. Fig. 4 shows the magnetization versus temperature (*M-T*) curves of samples measured at 50 Oe where *H* applied parallel to the *c*- axis of the samples. Both samples exhibit Pauli paramagnetic type behavior up to ~125 K and then decrease very slowly down to ~10 K. Then, a very sharp transition to diamagnetic phase was observed at T_c. In addition, it is concluded that abnormal behavior (kink) around 125 K may be related to the structural transformation. In addition, it is observed that the field magnetization increases at low temperatures. This phenomenon indicates the presence of magnetic impurities found by XRD and SEM.

Figure 2. SEM micrographs of samples

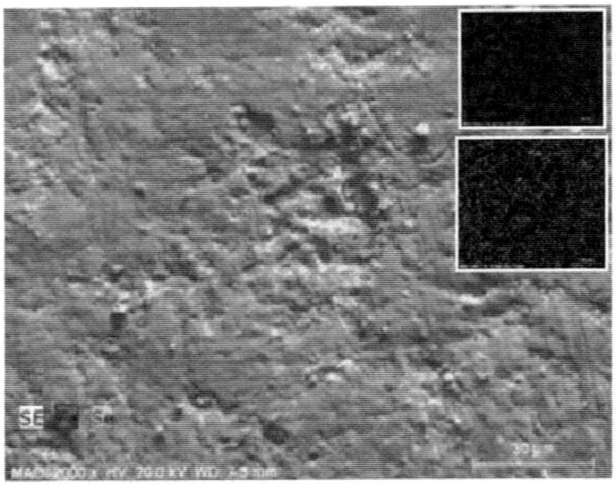

Figure 3. SEM micrographs of $FeSe_{0.98}$sample. Inserts show elemental distribution of Fe (red dots) and Se (yellow dots).

By looking at the magnetization results, it is observed that the superconducting fractions of samples are small exhibiting no full diamagnetic shielding. The magnetization versus magnetic field (*M-H*) loops of samples was given in Fig.5 for different temperatures. We have observed a similar shape of the *M-H* loops while temperature changed. It is seen that a narrow and ferromagnetic type *M-H* curves are formed. This suggests that the co-existence of both magnetism and superconductivity in our samples due to some impurity phase. The big difference on the magnetization scale is found, i.e. the magnetization value of $FeSe_{1.2}$ is about 5 times larger than those of $FeSe_{0.98}$. It is argued that the ferromagnetic components in $FeSe_{1.2}$ are dominant in the matrix and reveal an increasing at the magnetization values with increasing magnetic field.

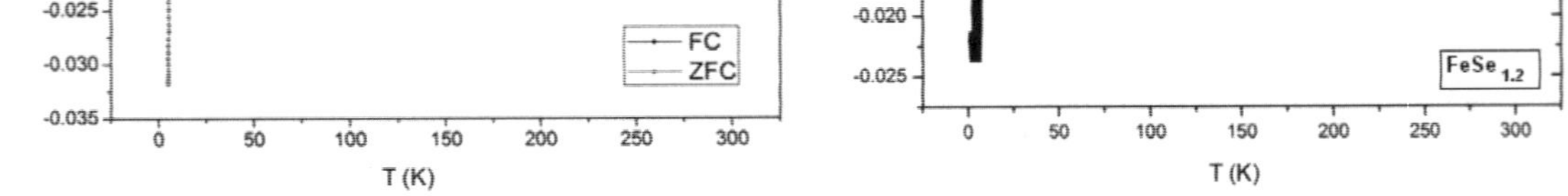

Figure 4. Magnetization versus temperature (M-T) curves of samples measured at 50 Oe, where *H* applied parallel to the *c*- axis of the samples.

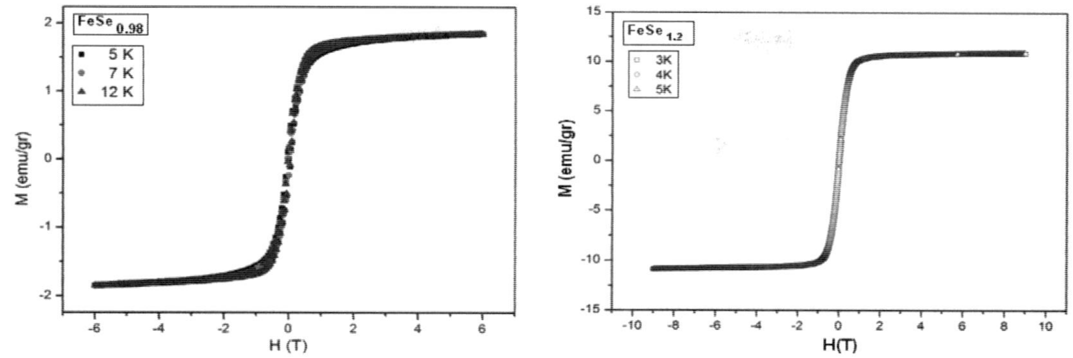

Figure 5. Magnetization versus magnetic field (M-H) loops of samples

4. Conclusions

The $FeSe_{0.98}$ and $FeSe_{1.2}$ bulk polycrystalline superconductors have been prepared by solid-state reaction method. Structural and magnetic properties were investigated. The XRD and SEM-EDX results exhibited that microstructural evolution of the samples was not strictly depending on the sample preparation process. However, in contrary, magnetic hysteresis measurements have shown some differences for the sample of $FeSe_{1.2}$. We have found that the ferromagnetic structure in $FeSe_{1.2}$ sample is dominant down to the low temperatures.

Acknowledgement

This work is supported by the TÜBİTAK (The Scientific and Technological Research Council of Turkey) under grant contracts no. 112T767 and Research Fund of Cukurova University, Adana, Turkey, under grant contracts no: FYL-2014-2363. The authors wish to thank Dr. Sabine Wurmehl, Michael Schulze and Prof. B. Büchner, from IFW Dresden (Germany) for their hospitality and support during the preparation of $FeSe_{0.98}$ sample.

References

[1] Hsu F C *et al* 2008 *Proc. Natl. Acad. Sci.,* **105** (38) 14262- 14264.
[2] Stewart G R *et al* 1984 *Rev. Mod. Phys.* **56** 755
[3] Kamihara Y *et al* 2008 *J. Am. Chem. Soc.* **130** 3296
[4] Ren Z A *et al* 2008 *Chin. Phys. Lett.* **25** 2215
[5] Margadonna S *et al* 2008 *Chem. Comm.* **5**607
[6] Pulikkotil J J *et al* 2008 *Supercond. Sci. Technol.* **23** 054012
[7] Williams A J *et al* 2009 *Solid State Communications* **149** 1507
[8] Li Y F *et al* arXiv 0905.3303.
[9] Lee K W *et al* 2008 *Phys. Rev. B* **78** 174502.
[10] Kotegawa H *et al* 2008 *J. Phys. Soc.Japan* **77** 113703.
[11] Masaki S *et al* 2009 J. *Phys. Soc. Jpn.* **78** 063704
[12] Yoshida R *et al* 2009 *J. Phys. Soc. Japan* **78** 034708
[13] Athena S S *et al* 2013 *Current Opinion in Solid State and Materials Science* **17** 59
[14] Diko P *et al* 2012 *Physica C* **476** 29
[15] Felner I *et al* 2008 arXiv *cond-mat* **0805** 2794
[16] Onar K *et al* 2015 *Journal of Alloys and Compounds* **620** 210
[17] Güler Kantarcı N *et al* 2014 *J. Of Superconductivity and Novel Magnetism* **27** 2691

The synthesis and magnetic structure of the iron selenide $Ba_{0.8}Fe_2Se_2$

N Kantarcı Güler[1], M Gürsul[1], C Işık[1], A Ekicibil[1], B Özçelik[1], H Okazaki[2], H Takeya[2], Y Takano[2]

[1]Department of Physics, Faculty of Sciences and Letters, Çukurova University, Adana, Turkey
[2]National Institute for Materials Science-1-2-1 Sengen, Tsukuba, Ibaraki 305-0047, Japan

E-mail: ozcelik@cu.edu.tr

Abstract. The $Ba_xFe_{2-y}Se_2$ (x=0.8) single crystal has been synthesized by using the one step technique at relatively lower temperatures. We have investigated structural properties of samples by using the XRD, SEM-EDX and magnetic properties. The SEM results clearly demonstrate that Ba ions are intercalated in the FeSe lattice. XRD result shows that the sample prepared has multi-phase structure. According to our *M-H* measurements perfect diamagnetism has been observed only in low field at 5 K. In addition, those curves indicate that the high field value and some impurities reveal ferromagnetic interactions. The superconducting transition was observed at around T_c≈10.9 K.

1. Introduction

After the discovery of superconductivity in iron chalcogenides, FeSe having the critical temperature of T_c = 8 K, with the layered anti *PbO*- tetragonal structure at room temperature was followed and activated considerable interest in studying these compounds in more detail because of their having the simplest crystal structure among various iron based superconductors [1]. Subsequently, alkali metal intercalated FeSe superconductor [2] around the transition temperature T_c≈30 K was reported and then, a chalcogenide analog to the 122 iron pnictide superconductors [3, 4] was found as very interesting topic for the researchers.

Later on the first announced synthesis [2] of $K_{0.8}Fe_2Se_2$, different compounds with the formula $A_xFe_{2-y}Se_2$ intercalated with A= Cs [5], Rb [6], Tl/K [7] and Tl/Rb [8] were successfully prepared. The critical temperatures of all these superconductors were reported around 30 K. $A_xFe_{2-y}Se_2$ compounds exhibit several distinctive features comparing to other iron-based superconductors: ***(i)*** Fe deficiencies in the FeSe layer, proximity to an anti-ferromagnetic semiconducting state [9], ***(ii)*** an extremely high Neel transition temperature [10], and ***(iii)*** the absence of hole pockets [11]. The barium transition metal chalcogenides $Ba_xM_yX_z$ (where M=3d metal, X= Se or S), compounds having different stoichiometries (x:y:z of 2:1:3, 1:1:2, 1:4:3, 1:2:2 and 1:2:3) could be synthesized [13]. For all these compounds, it is known that there are nine different structure types. It has been found that all these structures were based on columns of face-sharing BaX_6 tetrahedral. In Ba112 and Ba122, sheets are created by edge sharing of the neighboring columns. But, both in Ba123 and in Ba143, the columns are isolated [13]. In particular, Ba123 seems to be an interesting analog to K122, as the additional charge of the alkaline earth metal is balanced by the additional chalcogenide anion in the formula [14]. For growing of the single crystals of $A_xFe_ySe_2$, the Bridgman has typically been used. As it is well-known, this synthesis technique is performed at high temperatures above the melting point of the reactants and through a long growth process. In this work, we have synthesized the $Ba_xFe_{2-y}Se_2$ sample by using the one-step-method at lower temperatures [12,22] and investigated the structural properties by using the XRD, SEM- EDX and magnetic properties with *M-T* and *M-H* measurements under the low and high magnetic fields. We have also calculated the critical current density, J_c by using Bean Critical State Model formula [20]. The critical current density data have been used to obtain the volume pinning force F_p, originating from the interactions between the vortex lattices and pinning defects per unit volume in the $Ba_xFe_{2-y}Se_2$ single crystal.

2. Experimental procedure

$Ba_{0.8}Fe_2Se_2$ was synthesized by a one step technique. First, BaSe (99.9%), Fe (99%) powders and Se grains (99.999%) were put into an alumina crucible and sealed into an evacuated quartz tube. All preparation steps were carried out in argon filled glove box with O_2-H_2O content less than 0.1 ppm. The quartz tube was heated to 930 °C and kept constant for 12 hours. Following this, the tube was cooled down to 550 °C temperature followed by quenching to room temperature. The obtained samples were characterized by Scanning Electron Microscopy (SEM) together with energy dispersive X-ray (EDX) spectrometer (LEO Evo-40 VPX) , X-ray diffraction (XRD) (Rigaku D/max-B powder diffractometer system working with CuKα radiation), magnetization and magnetic hysteresis techniques (PPMS, Quantum Design magnetometer).

3. Results and discussions

The X-ray powder diffraction patterns of $BaFe_{2-y}Se_2$ sample is presented in Fig.1. Among the XRD patterns the six strong peaks appear at about $2\theta \approx 24.638$, 29.533, 48.276, 50.376, 52.129 and 76.652 degrees. These peaks correspond to the reflected intensity from the (202), (311), (321), (323), (024), (307) planes of the orthorhombic structure having space group *Pnma* with the lattice parameters of $a = 11.85(4)$ Å, $b = 5.47(1)$ Å, $c = 9.15(8)$ Å and $V = 593.92(5)$ Å^3 [15]. Calculation of the unit cell parameters were done by using Jade 6.0+ crystal refinement program.

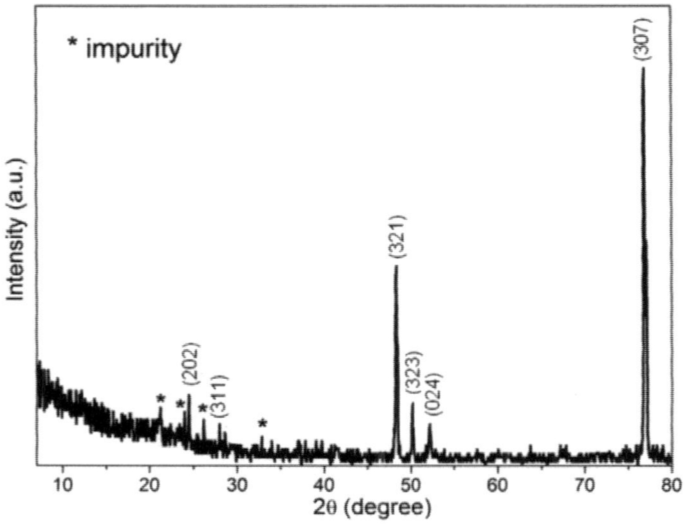

Figure 1. X-ray diffraction patterns of the $Ba_{0.8}Fe_2Se_2$ sample

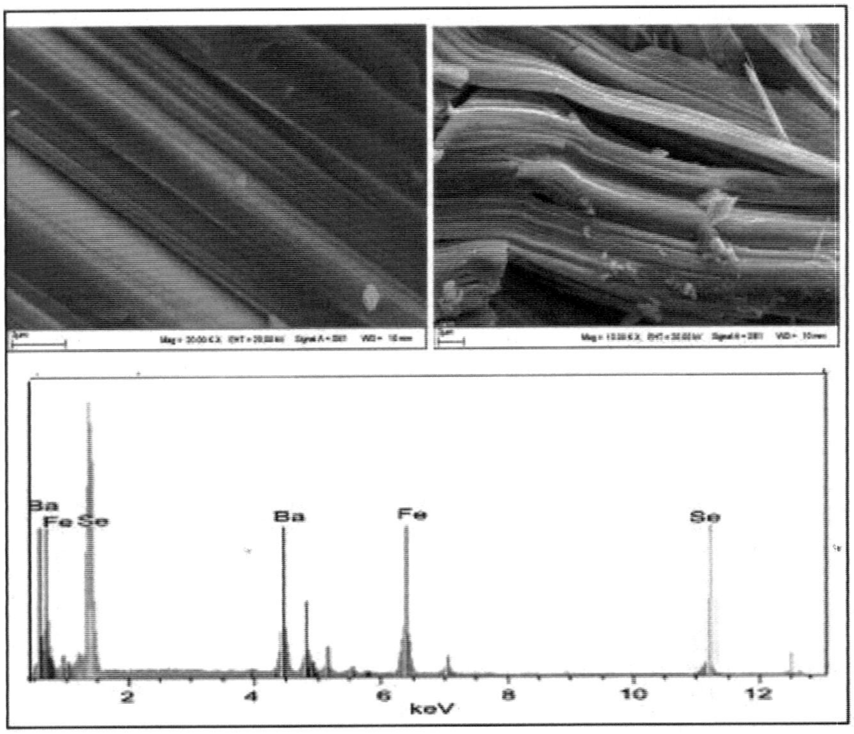

Figure 2. SEM images and EDX spectrums of the $Ba_{0.8}Fe_2Se_2$ sample

The uncertainty of the calculation remained in the ±0.0001 range. In addition to the above mentioned intensity peaks, sample has also shown some very weak additional peaks corresponding to impurity phase. These new impurity peaks in the XRD spectra may be due to different Fe-Se phases (Fe3Se4 or Fe7Se8) remains in the compound during the formation of BaFeSe. It may be concluded that the solubility of the Fe in the BaFeSe compound has been decreased by the annealing process. These results are in agreement with literature [14,15].

The SEM images 20.000 and 10.000 times magnified are shown in Fig 2. It is quite clear that the surface morphology of the sample is the terrace like formations together with well connected small rectangular bars. It may be argued that these structures are responsible from superconductivity. Fig. 2 also shows the corresponding energy dispersive x-ray (EDX) spectra of the $Ba_{0.8}Fe_2Se_2$ sample. From the similarity of the Fe, Ba and Se peak intensity line traces in the EDX spectra of the sample, it is concluded that after the synthesis process, Fe, Ba and Se have been almost homogeneously distributed throughout the sample. All the locations in the matrix contain the expected elements (Fe, Ba, Se) pointing out no other impurity elements. In addition, the nominal chemical compositions of the sample determined by the EDX analyses are $Ba_{0.69}Fe_{1.83}Se_2$. Namely, the EDX line traces indicate that Ba element integrated into the structure of FeSe. In addition it is said that instead Ba122 structure, almost Ba123 one has formed.

The magnetic properties of the $Ba_{0.8}Fe_2Se_2$ sample have been also investigated. The temperature dependences of zero field cooled (ZFC) and field cooled (FC) magnetization at 10 Oe for sample are shown in Fig. 3. Above T_c sample has shown a Pauli paramagnetic type behavior without changing down to T_c. But a sharp drop to diamagnetic phase was obtained after T_c. The zero field cooled (ZFC) magnetization shows that the superconducting shield begins to emerge at superconducting transition temperature and then show a clear transition.

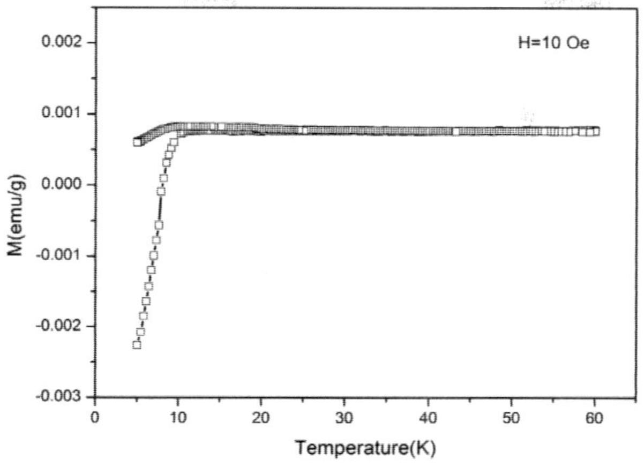

Figure 3. FC and ZFC magnetizations aganist temperature for 10 Oe

The sample shows onset of diamagnetism at temperature, T_c^{mag}, 10.9 K. As seen in figure, the field cooled (FC) magnetization is very small below the T_c^{mag} temperature for the sample. This is not imply the superconducting volume fraction is very small (around <3%) , since the diamagnetic signal at FC mode is very small even in $Ba_{0.8}Fe_2Se_2$ where bulk superconductivity is confirmed to exist from the specific heat measurements, as reported in previous papers [16,17]. This is probably due to the strong vortex-pinning effect in $Ba_{0.8}Fe_2Se_2$ [18]. In addition, this extremely small magnetic moment in the FC measurements is possibly a result of the local lattice distortions caused by the substitution with relatively large ionic radius of Ba ions, introducing effective pinning centers [19].

The two types of magnetic hysteresis measurements were performed on the samples: the magnetizations as a function of low and high magnetic fields for different temperatures. The results are plotted in Figs. 4 and 5.

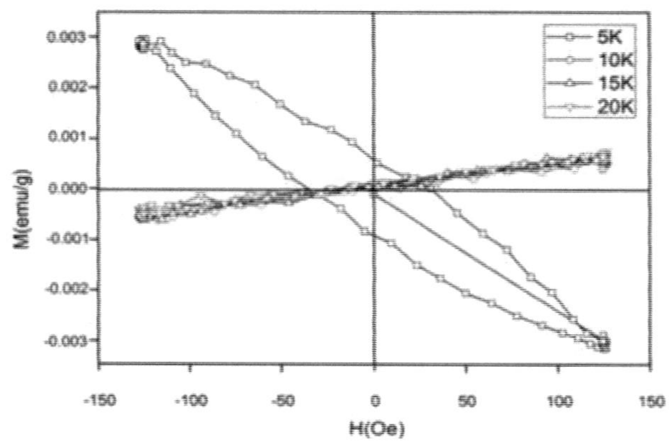

Figure 4. Magnetization against low values of applied field for different temperatures

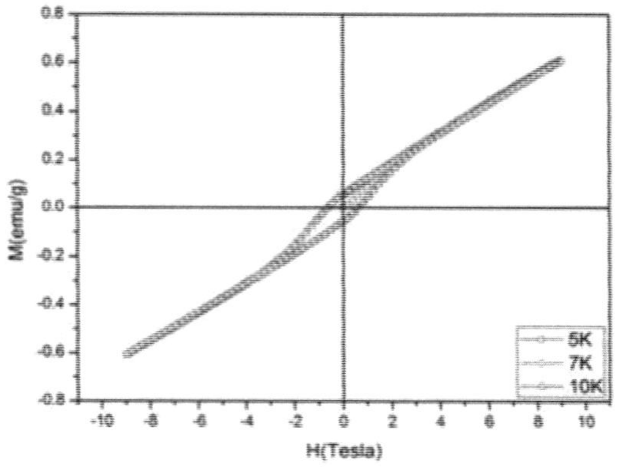

Figure 5. Magnetization against high of applied field for different temperatures

The M-H curves have revealed the existence of ferromagnetism as seen in Fig. 4 for temperatures 10, 15 and 20 K. The small amount of impurities may be the reason of the ferromagnetic components in *M-H* curves. It further supports the fact that the weaker coupling between each plane reveals in more layer-like transport characteristics as the distance between neighboring Fe-occupied planes increases. As can be seen from Fig. 4, the superconducting behavior is found at the lower magnetic fields for the temperature of 5 K. At this temperature, the sample behaves a type-II superconductor and shows nearly perfect diamagnetism. The area of the *M-H* loops indicate the ferromagnetic (nearly paramagnetic above $\approx$ H > 3 Tesla) behavior in the high magnetic fields (see Fig 5). This suggests that the co-existence of both magnetism and superconductivity in our sample due to ferromagnetic nature of Fe and impurity phase, as seen in the XRD analysis.

The critical current at 5 K for the lower magnetic fields can be obtained from the magnetic hysteresis curve by using the Bean's Critical State Model [20] given as:

$$J_c = 20\Delta M/(a(1\text{-}a/3b)) \qquad (1)$$

where $\Delta M = M+ - M-$ is measured in electromagnetic units per cubic centimeter, a and b are the dimensions of the cross section of the sample perpendicular to the applied field. We have estimated the critical current density of the sample at temperatures of 5K, and plotted in Fig.6 against applied magnetic field for the fields in the range $0 \leq H \leq 130$ Oe.

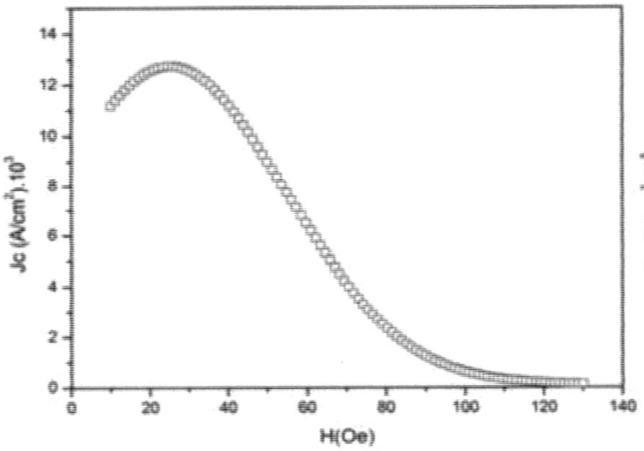

Figure 6. Calculated critical current density against low magnetic field at 5K

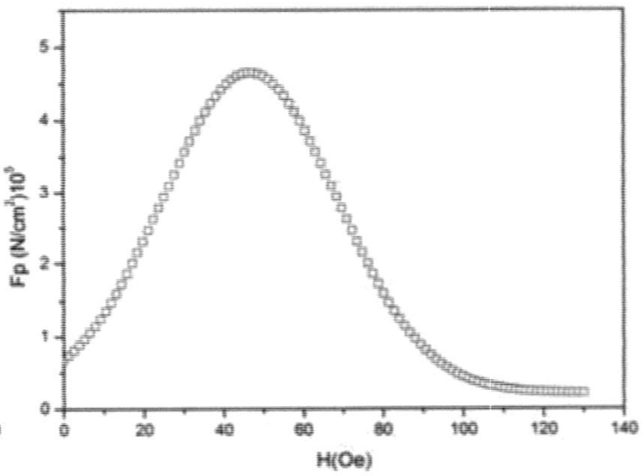

Figure 7. Variation of pinning forces against applied low magnetic fields

It is clear that J_c increases in low fields region, and then decreases in high fields. Generally, J_c is composed by inter-granular and intra-granular parts of crystal [21]. As can be understood, the weak links exist for inter-granular. In case of existence of large numbers of weak links, for higher fields, J_c from the inter-granular part should be much more sensitive to magnetic field, which thus causes a decrement at J_c. The critical current density data have been used to obtain the volume pinning force F_p, originating from the interactions between the vortex lattices and pinning defects per unit volume in the $Ba_{0.8}Fe_2Se_2$ sample, studied at temperature 5 K by using the formula:

$$\boldsymbol{F_p = J_c \, x \, B} \qquad (2)$$

The plot of F_p versus H is given in Fig.7. The maximum value of F_p is found around 50 Oe.

4. Conclusions

We have systemically studied the structure and magnetic properties of $Ba_{0.8}Fe_2Se_2$. From XRD patterns multiphase structure was found. SEM images show that the surface morphology of the sample is the terrace like formations together with well connected small rectangular bars. The nominal stoichiometries of Ba, Fe and Se in the sample are determined as 0.69, 1.83, 2, respectively. This indicates that the Ba123 structure has formed in the matrix together with a small part of superconducting volume fraction (around <3%). The diamagnetic onset temperature obtained from the ZFC magnetization is 10.9 K. J_c value calculated from low field *M-H* curve measured at 5 K decreases by increasing the applied magnetic field, suggesting the weak-link behavior in the sample.

Acknowledgement

This work is supported by the TÜBİTAK (The Scientific and Technological Research Council of Turkey) under grant contracts no. 112T767. We wish to thank Dr. Masashi Tanaka from National Institute for Materials Science (NIMS)-Japan for valuable discussions.

References

[1] Hsu F C *et al* 2008 *Proc. Natl. Acad. Sci.* USA **105** (2008) 14262
[2] Guo J *et al* 2010 *Phys. Rev. B* **82** (2010) 180520(R)
[3] Rotter M *et al* 2008 *Phys. Rev. Lett.* **101** (2008) 107006
[4] Sasmal K *et al* 2008 *Phys. Rev. Lett.* **101** (2008) 107007
[5] Krzton Maziopa A *et al* 2011 *J. Phys. Condens. Matter* **23** (2011) 052203
[6] Wang A F *et al* 2011 *Phys. Rev. B* **83** (2011) 060512
[7] Fang M *et al* 2011 *Europhys. Lett.* **94** (2011) 27009
[8] Wang H D *et al* 2011 *Europhys. Lett.* **93** (2011) 47004
[9] Fang M H *et al* 2011 *EPL* **94** (2011) 27009
[10] Bao W *et al* 2011*Chin. Phys. Lett.* **28** (2011) 086104
[11] Zhang Y *et al* 2011 *Nat. Mater.* **10** (2011) 273
[12] Ozaki T *et al* 2012 *EPL* **98** (2012) 27002
[13] Iglesias J E *et al* 1975 *Kristallogr.* **142** (1975) 398
[14] Krzton Maziopa A *et al* 2011 *J. Phys.: Condens. Matter* **23** (2011) 40220
[15] Svitlyk V *et al* 2013 *J. Phys.: Condens. Matter* **25** (2013) 315403
[16] Noji T *et al* 2010 *J. Phys. Soc. Jpn.* **79** (2010) 084711
[17] Noji T *et al* 2012 *J. Phys. Soc. Jpn.* **81** (2012) 054708
[18] Ohno M *et al* 2014 *J. Phys. Soc. Jpn.* **83** (2014) 044704
[19] Yeh K W *et al* 2009 *Crys Growth Desing,* **9** (2009) 4847-4851
[20] Bean C P *et al* 1964 *Rev. Mod. Phys.* **36** (1964) 31
[21] Zhuang J C *et al* 2013 *Low Temp. Phys.* **172** (2013) 113
[22] Onar K *et al* 2015 *Journal of Alloys and Compounds* **620** 210

Microcavity properties of 2D photonic crystal made by silica matrix doped with magnetic nanoparticles

R Moukhtari, A Hocini and D Khedrouche

Laboratoire d'Analyse des Signaux et Systèmes
Department of Electronics, University of Mohamed Boudiaf of M'sila
BP.166, Route Ichebilia, M'sila, 28000 Algeria

E-mail: hocini74@yahoo.fr

Abstract. In this present paper, quality factor of two-dimensional magneto-photonic crystals microcavity fabricated by SiO2/ZrO2 or SiO2/TiO2 matrix doped with magnetic nanoparticles, in which the refractive index varied in the range of 1.51 to 1.58, has been investigated. Finite difference time domain method (3D FDTD) with perfectly matched layers (PML) was used to calculate the transmission spectrum. We demonstrate that the Q factor for the designed cavity increases as the refractive index increases, and found that the Q factor decreases as the volume fraction VF% increases. The obtained results are useful for better designs of magneto photonic crystal devices.

1. Introduction

Recently, optical properties of sol gel process have been intensively studied [1–4], because of their importance in the manufacturing of optical devices such as optical isolator and circulator. A sol–gel process, a promising method to prepare metal oxide films at low temperature, is an attractive alternative to the other deposition techniques [5–6]. This technique can be used through simple procedures in the atmosphere and an atmospheric pressure, and it realizes the nanometer level control of thickness of each layer similar to dry processes. In addition, it has some advantages for the preparation of PCs with magneto-optical function, i.e., strong interlayer coupling, relatively low environmental burden and low capital cost. However, there have been scarce reports on preparation of PCs applicable to the magneto-optics using the sol–gel method [7].

Magneto-photonic crystals (MPCs) are a special kind of photonic crystals (PCs), constructed of magnetic materials. The first report arrived at the end of the 1990s, and at the beginning of the new millennium activity in this area is growing rapidly. The research on two-dimensional (2D) MPCs has been undertaken recently, together with expansion of the capability for a new class of devices more suitable for integration. This includes planar nonreciprocal devices specific to 2D MPCs, such as optical isolators based on line-defect waveguides [8] and circulators based on micro cavities [9], which have been demonstrated theoretically. The study of optical cavities based on magneto-photonic crystal slabs has attracted much attention [10–11]. There are many possible device applications of compact and efficient (MPCs) cavities, such as magneto-optical resonators, circulators, etc. In this case, the design of magneto-photonic crystal cavity with high quality factor and prescribed frequency response is highly needed for these applications. In this paper, we theoretically investigated the quality factor of magneto-photonic crystal microcavities, made by SiO2/ZrO2 or SiO2/TiO2 matrix doped with magnetic nanoparticles with low refractive index

Journal of Physics: Conference Series **667** (2016) 012004 doi:10.1088/1742-6596/667/1/012004

material which varied in the range of 1.51 to 1.58 at λ =1.55 μm [12], using the finite difference time domain 3D (FDTD) method assisted by group theory in the time domain [13]. The magneto-photonic crystal microcavities (H1) are obtained by removing one hole missed of the 2D magneto-photonic crystal structure. We simulate a cavity that the MPC structure to ensure high transmission efficiency. In the first part of this work, we have investigated the effects of the contrast on the *Q* factor. In the second part, we have varied the concentration of magnetic nanoparticles VF% of the structure: magneto-photonic cavity realized by sol–gel process, and we study the influence of volume fraction VF % on the position of resonance wavelength and Q value.

2. Theoretical model

The model is a MPCS composed of 2D triangular lattice magneto-photonic crystal of air holes as shown in Figure 1. However, the triangular lattice is very suitable to study photonic band gap properties. Triangular lattice allows the opening of 2D photonic band gap, presents a good compromise namely for high felling factors and its lowly sensitive to the incidence angle compared to the square lattice [14-15]. In this paper, the lattice constant is 0.75 μm, which is a distance between the two neighbored holes, given as 'a'. The radius of the hole is 0.27 μm. This structure is realized by SiO2/TiO2 or SiO2/ZrO2 material with low index of refraction which varied in the range of 1.51 to 1.58 at λ =1.55μm [16]. The composite matrix is created using a doping procedure where the magnetic nanoparticles are introduced in the liquid preparation of a sol-gel process. SiO2/TiO2 or SiO2/ZrO2 slab have been doped with Maghemite nano particles whose average diameter is about 10 nm [17-18]. The magneto-optical permittivity tensor $\hat{\varepsilon}$ of slab is written in coordinate system as:

$$\hat{\varepsilon} = \begin{bmatrix} \varepsilon & \varepsilon_{xy} & 0 \\ -\varepsilon_{xy} & \varepsilon & 0 \\ 0 & 0 & \varepsilon \end{bmatrix} \tag{1}$$

Where ε represents the diagonal elements of dielectric tensor ε_{xx},ε_{yy} and ε_{zz}. In our case, these elements are equal and they are given as follows:

$$\varepsilon = \varepsilon_{xx} = \varepsilon_{yy} = \varepsilon_{zz} = \varepsilon^{'} + \varepsilon^{''} \tag{2}$$

However, the complex refractive index of magneto optical matrix is given by: $N = n + ik$, where n is the real part of complex refractive index for diagonal elements tensor and k is the imaginary part of complex refractive index. The complex dielectric permittivity of diagonal elements is given by:

$$\varepsilon = (n + ik)^2 \tag{3}$$

The real and the imaginary parts of dielectric tensor are given by:

$$\begin{cases} \varepsilon^{'} = n^2 - k^2 \\ \varepsilon^{''} = 2nk \end{cases} \tag{4}$$

where k is called the coefficient of extinction and it is given by:

$$k = \frac{\alpha.\lambda}{4\pi} \tag{5}$$

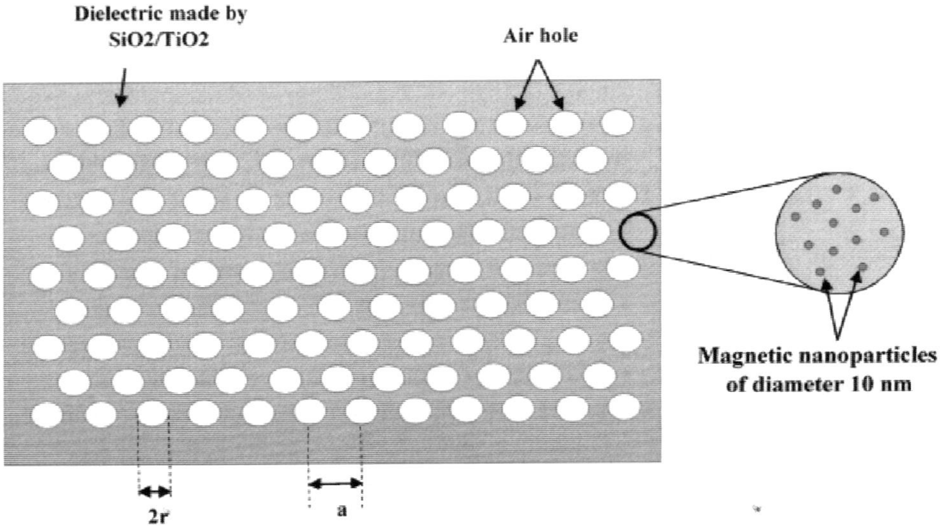

Figure 1. Structure of 2D photonic crystal made with SiO2/TiO2 or SiO2/ZrO2 doped with magnetic nanoparticles.

Where λ is the free space wavelength and α (/cm) represents losses inside the matrix, it is linked to intrinsic absorption of magneto-optical matrix. However, the application of a magnetic field on material with a direction parallel to the light beam (Oz) produces an off-diagonal elements where the magnitude depends on the kind of material (on the faraday rotation θ_F and ellipticity η_F) [19].

$$\varepsilon_{xy} = -\varepsilon_{yx} = \varepsilon_{xy}^{'} + \varepsilon_{xy}^{''} \tag{6}$$

The real and the imaginary parts of ε_{xy} are strongly depending on the Faraday rotation θ_F that is proportional to the concentration of magnetic nanoparticles VF%. These elements are given by:

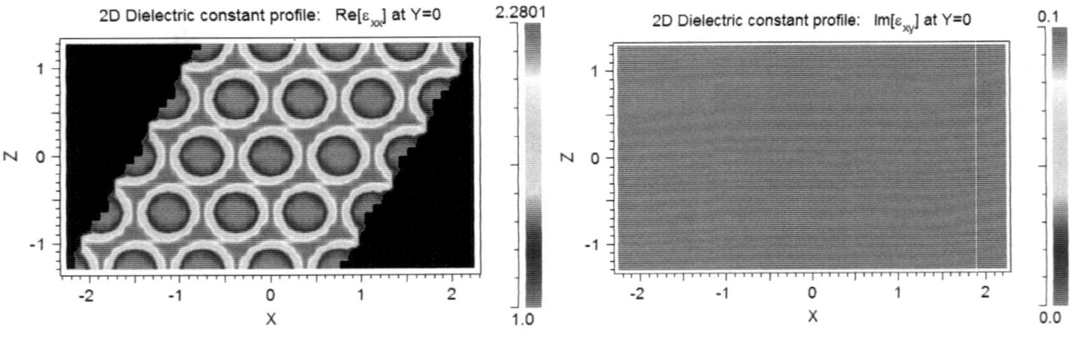

Figure 2. The dielectric constant: diagonal dielectric constant $\varepsilon = \varepsilon_{xx} = \varepsilon_{yy} = \varepsilon_{zz} = 1.51^2$, and off-diagonal dielectric constant $\varepsilon_{xy} = -\varepsilon_{yx} = 0.016^2$.

$$\begin{cases} \varepsilon'_{xy} = \frac{\lambda}{\pi}(n\theta_F - k\eta_F) \\ \varepsilon''_{xy} = \frac{\lambda}{\pi}(n\theta_F + k\eta_F) \end{cases} \tag{7}$$

In the limit of small off-diagonal components: $|\varepsilon_{xy}| << |\varepsilon|$ and small absorption, an approximate expression for the imaginary part of ε''_{xy} is obtained [19–20]:

$$\varepsilon''_{xy} = \frac{\theta_F . \lambda\sqrt{\varepsilon}}{\pi} \tag{8}$$

Figure 3, shows the band diagrams of non-magnetic structure of 2D photonic crystal for TE (blue) and TM (red) modes of the polarization. The matrix have a permittivity $\varepsilon = 1.56^2$ and the magnitude of the off-diagonal elements are $\varepsilon_{xy} = 0$, corresponding to the SiO2/TiO2 or SiO2/ZrO2 at optical communication wavelength λ=1.55 µm [21–22]. A dispersion diagram showing normalized frequency versus the wave vector for transverse electric (TE) and transverse magnetic (TM) modes of the 2D photonic crystal is given in Figure 2. It has been calculated along the Γ-X-M-Γ edge for the Brillouin zone by employing a 2D plane wave expansion (PWE) method. The gap map show two frequency band gaps for the TM polarized modes but no gap for TE modes with this refractive index contrast and relatively small r/a. In the TM band diagram the fundamental band gap is centered near a wavelength of 1.5506. It extends between to wavelength range 1.5052-1.5961µm for the waves with TE polarization for the refractive index n=1.56.

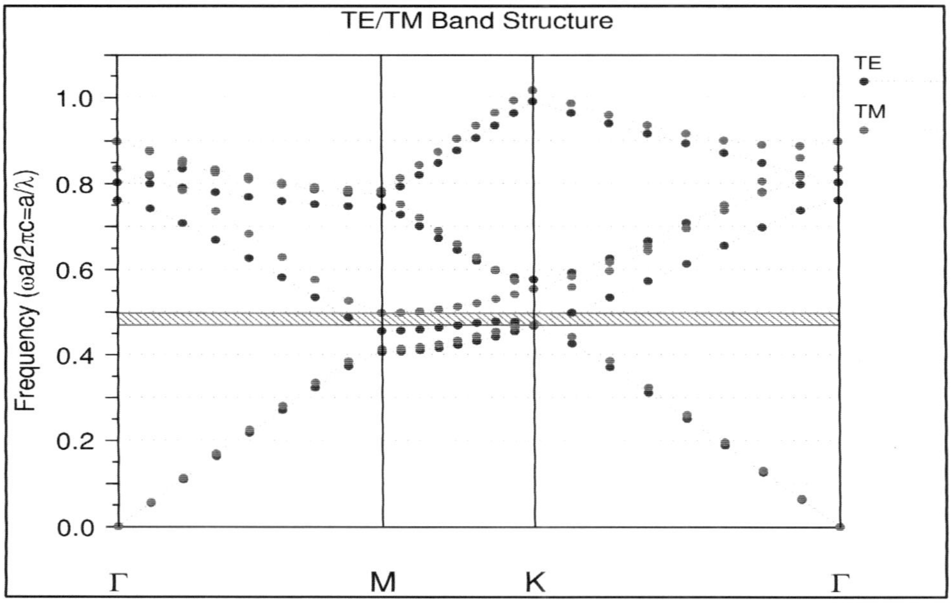

Figure 3. Dispersions curves and band-gaps for TM and TE polarizations for the 2D lattice without defects.

3. The microcavity design

This section presents the influence of refractive index contrast, was changed in order to predict the highest quality factor Q to improve the transmission efficiency. The numerical tools used for our simulations are based on the two-dimensional finite difference time-domain method (FDTD) (RSoft Design Group, FullWAVE) [23]. The structure of our 2D PC cavity is shown in Figure. 4, with one missing hole in the center, where the number of circular holes considered for both 'X' and 'Z' directions are 11. This cavity allows for very high quality factors while maintaining a size of the order of optical wavelength.

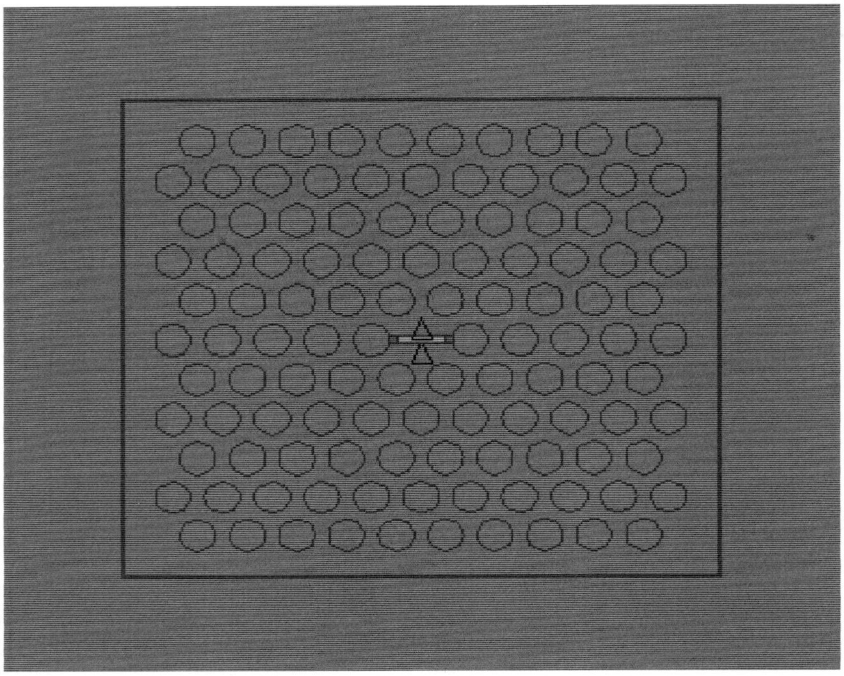

Figure 4. Sight of top of the H1 cavity made by removed one hole.

We begin our consideration by calculation the transmission spectra for a variety of refractive index (n) in the range 1.52 to 1.58. Figure 5. shows the spectral response of the microcavity formed by removing one hole obtained with the FDTD method of the impulse response for different values of refractive index n = 1.52 to 1.58. The spectrum calculated by TM polarization can be observed one cavity mode for different values of refractive index. We will call the mode of mode resonance wavelength. In order to show the location of resonance wavelength and quality factor, Fullwave software is used to simulate the transmission spectrum for different values of refractive index n = 1.52, to 1.58 .

Figure 5 reports the simulations. The results show that more refractive index increase the resonance wavelength shifts towards the right $\lambda 0$ = 1.5225 to 1.5465μm, and quality factor Q= ($\lambda 0/\Delta\lambda$) varied from Q = 57 to 116. Figure 6. Presents the variations of resonance wavelength and quality factor as a function of refractive index. We begin our calculation of transmission spectra by increasing the refractive index. We show the dependence of quality factor and resonance wavelength as a function of the symmetric than the refractive index increases. After scanning of refractive index n carefully, we find the highest Q value Q = 116 with resonant frequency 1.5465 μm, which can be achieved at n = 1.58.

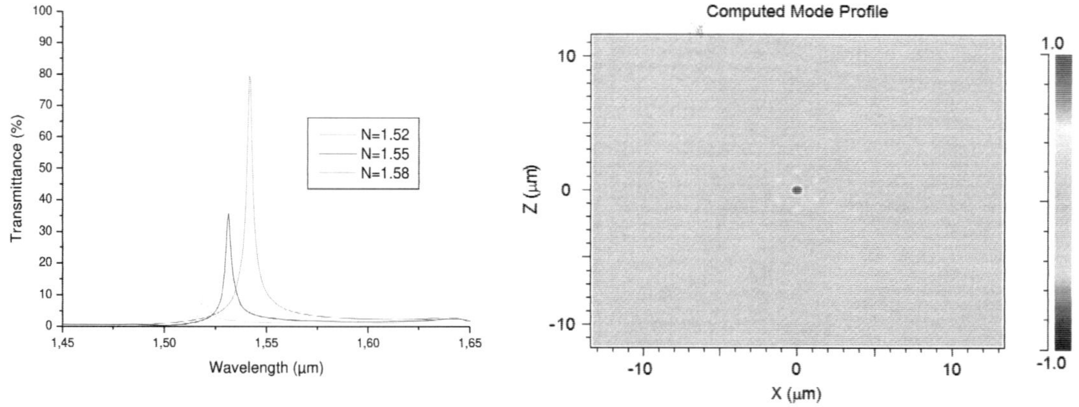

Figure 5. (a) Frequency response of the H1 microcavity formed by removing one hole triangular lattice as a function of refractive index, (b) the intensity field distribution for the microcavity at resonant wavelength 1.5465 µm (n=1.58).

4. Effect of the volume fraction

The refractive index of the structure was fixed at n = 1.58 and at series of simulations for different values of volume fraction VF%, from VF = 1% (n_{xy} =0.016) to 20% (n_{xy} = 0.073). We begin our calculation of transmission spectra in Figure 7 by increasing the volume fraction. In addition, in Figure 8, we show the dependence of quality factor and resonance wavelength as a function of increased the volume fraction. The quality factor decreases as the volume fraction increased. After scanning of VF% carefully, we find the final Q value $Q = 21$ with resonant frequency 1.5506µm, that can be achieved at VF = 20% (see Figure 7 and Figure 8).

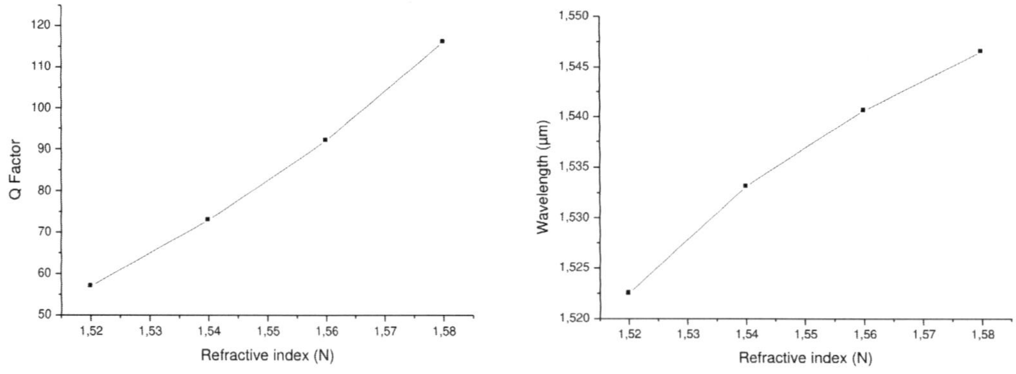

Figure 6. Change in quality factor and Wavelength as a function of refractive index of the H1 cavity.

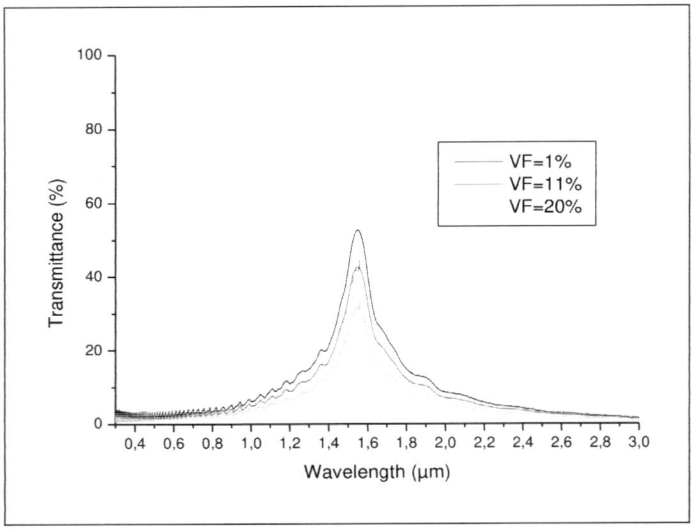

Figure 7. Frequency response of the H1 microcavity as a function of volume fraction (VF %).

The computational method used is based on a 3D finite difference time domain (3DFDTD) method algorithm. Perfectly, matched layer (PML) conditions have been considered in the calculations to ensure no back reflection in the limit of the analyzed region. This crystal is light by a Gaussian wave under normal incidence with a transverse electric (TM) polarized. The time step is chosen to 0.01. Note that it might be necessary to reduce the time step below the stability limit when simulating metals since the courant condition can change in this case.

To explore the effect of volume fraction due to off-diagonal elements, we have calculated the quality factor of MPC cavity for different values of volume fraction (VF %). We show in Figure 8, variations of quality factor and resonance wavelength as a function of the increased volume fraction of the MPC structure. From figure 4, we notice that the volume fraction due to off-diagonal elements have a great effect on the position of resonance wavelength and Q factor. The quality factor Q decreases from 68 for 1% to 21 for 20%. It is clear that as the VF% increases, the quality factor Q decreases, this leads to a weakened in transmission spectra.

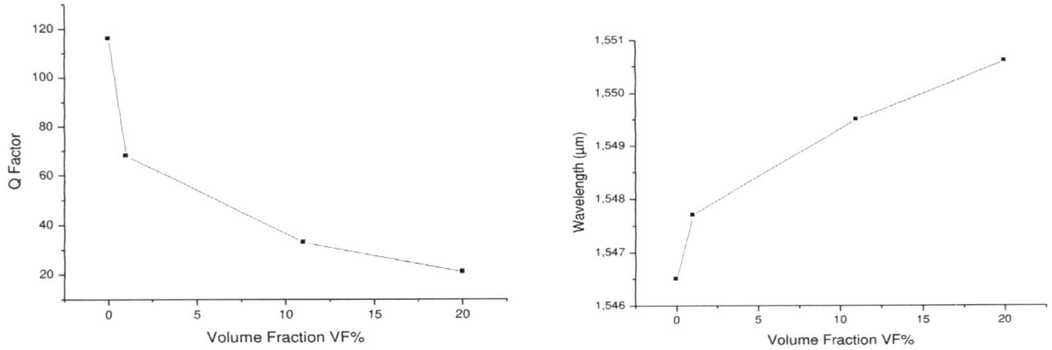

Figure 6. Change in quality factor and Wavelength as a function of volume fraction of the MPCH1 cavity.

5. Conclusion

In summary, a novel magneto-photonic crystal platform cavity has been studied in this paper. The microcavity made by SiO2 / ZrO2 or SiO2 / TiO2 matrix doped with magnetic nanoparticles, characterized by low refractive index material which varied in the range of 1.51 to 1.58. In addition, we have numerically study the effect of the volume fraction due to off-diagonal elements on the Q factor for the H1 cavity in a 2D magneto-photonic crystal.

The different results of simulation obtained by 3D FDTD method in Fullwave software demonstrate clearly that the volume fraction due to off-diagonal elements have a great effect on the position of resonance wavelength and Q value. Thus, we reveal that a maximum absolute quality factor is achieved for non-magnetic structure VF=0% (n=1.58), and VF=1% for magneto-photonic crystal made by sol–gel process.

References

[1] Hocini A, Boumaza T, Bouchemat M, Royer F, Jamonc D, and Rousseau J J 2008 *J. Microelectronics* **39** 99–102.
[2] Hocini A, Boumaza T, Bouchemat M,Choueikani F, Royer F, and Rousseau J J 2010 *J. Appl. Phys. B* **99** 553–558.
[3] Hocini A, Bouras M, and Amata H 2013 *J. Optical Materials* **35** 1669–1674.
[4] Hocini A, Bouchelaghem A, Saigaa D, Bouras M, Boumaza T, and Bouchemat M 2013 *J. Comput. Electron.* **12** 50–55.
[5] Mening M, Oliveira P W, and Schmidt H 1999 *J. Thin Solid Films.* **35** **1**99.
[6] Almeida R M, and Rodrigues A S 2003 *J. Non-Cryst. Solids* **326** 405.
[7] Inui C, Tsuge Y, Kura H, Fujihara S, Shiratori S, and Sato T 2008 *J. Thin Solid Films* **516** 2454–2459.
[8] Otmani H, Bouchemat M, Hocini A, and Boumaza T 2014 *J. Phys. Scr.* **89** 065501.
[9] Jin X, Ouyang Z, Wang O, Lin M, Wen G, and Wang J 2014 *J. PLOS ONE.* 0113508.
[10] Wang Z, and Fan S 2005 *J. Appl. Phys. B* **81** 369–375.
[11] Wang Z, and Fan S 2005 *J. Optics Letters* **3** 1989–1991.
[12] Kahlouche A, Hocini A and Khedrouche D 2014 *J. Comput. Electron* **6** 10825.
[13] Okano M, and Noda S 2004 *J. Phys. Rev. B.* **70** 125105.
[14] Kuang W, Hou Z, Liu Y and Li H2013 *J. Opt. A.* **7**525–528.
[15] Beghoul M R, Boudrioua A, Kremer R, and Fontana M2006*J. Microelectronic Engeneering*
[16] Amata H, Royer F, Choueikani F, Jamon D, Broquin J E, Plenet J C, and Rousseau J J 2010 *J. SPIE* **7719**.
[17] Choueikani F, Royer F, Jamon D, Siblini A, Rousseau J J, Neveu S, and Charara J 2009 *J. Applied Physics letters* **94** 051113.
[18] Royer F, Jamon, D, Rousseau J J, Roux H, Cabuil V, and Zins D 2005 *J. Applied Pysics Letters* **86** 011107.
[19] Wittekoek S, Popma T J A, Robertson J M., and Bongers P F 1975 *J. Phys. Rev. B.* **12** 2777–2788.
[20] Dionne G F, and Allen G A 1993 *J. Appl. Phys.***73** 6127.
[21] Choueikani F, Royer F, Douadi S, Skora A, Jamon D, Blanc D, and Siblini A 2009 *J. Appl. Phys.* **47** 30401.
[22] Kekesi R, Royer F, Jamona D, Blanc-Mignon M F, Abou-Diwan E, Chatelon J P, Niveu S, Tombacz E 2013 *J. Appl. Phys.* **73** 6127.
[23] RSoft Design Group *FullWAVE*.

Synthesis of Fe_3O_4 nanoparticles for biomedical applications

K. Onar[1] and M. E. Yakinci[1,2]

[1]İnönü Üniversitesi, Fen Edebiyat Fakültesi, Fizik Bölümü, 44280-Malatya, Turkey
[2]Inönü Üniversitesi, Mühendislik Fakültesi., Biyomedikal Mühendisliği Bölümü, 44280-Malatya, Turkey

E-mail: kubra.onar@inonu.edu.tr

Abstract. Nano particles of Fe_3O_4 with the range of 6.9 and 14.2 nm. By using chemical coprecipitation method. Particle size effect according to magnetic properties were investigated together with the constant reaction and crystallization temperature. According to calculations peaks were indexed as cubic Fe_3O_4 for all samples. The hysteresis loops shows superparamagnetic nature. It was found that when Fe^{2+}/Fe^{3+} ratio increase 50% the particle size increases approximately one fold. We also realized that chemical coprecipitation method is suitable for biomedical applications of Fe_3O_4nano particles.

1.Introduction

Since fully exterminate of cancer cell is indispensable for accomplished treatment, total excision is the treatment of choice if possible. Depending on the location and the involvement of the tumor with surrounding tissues, surgery may not always be possible. Under these conditions, chemotherapy or radiotherapy becomes necessary. However, both therapy methods have been reported that have severe complications during treatment. They may harmful for the intact cells can be destroying them after several seance. Therefore the development of new techniques is inevitable. At the same time that could selectively carry molecules of the drugs to the diseased area without giving harm to the healthy tissues.

Recently developed nano medical technologies are suitable candidates particularly for cancer research. It has been easily used for examining and treating cancer cells because the excellent characteristics of nano scale particles may improve the medical effect of traditional chemotherapy or radiotherapy methodologies largely. Due to the high biosafety, easy availability, controllable and examinable characteristics according to other nanoparticles, magnetic iron oxide (Fe_xO_y) have been widely used in clinical applications in last decade [1-5]. For immunoassays, involving the use of magnetic nano reagents (MNR), which consist of bioprobe-coated MNP and liquid solvents, several magnetic technologies have been developed to improve the performance relative to the performance of current clinical methods such as the enzyme-linked immunosorbent assay (ELISA) and others [1]. Today many formation and fabrication methods have been examined to obtained sub-nano magnetic particles. such as micro-emulsion, and poly-process [6-8] and especially chemical coprecipitation, which is increasingly popular due to its simplicity, inexpensive equipment and potential for large scale production [9]. Particularly, particle size of 2 - 15 nm is found to be the best for clinical applications, mainly for cancer treatments. However, methods of fabrication are generally getting difficult and onerous and needs too much attention to obtain physicallyand magnetically in good properties. In this work, we have focused on the fabrication of nano particles and effect of particle size on the physical, magnetic and electrical properties and results obtained are reported.

2.Experimental

In this work the Fe_3O_4 nanoparticles were prepared by chemical coprecipitation method similar to previously obtained research groups [9-11]. The starting materials were $FeCl_3{\cdot}6H_2O$, $FeCl_2{\cdot}4H_2O$ and $NH_3{\cdot}H_2O$ (Alfa Aesar). A ferric salt solution and a ferrous salt solution were prepared from $FeCl_3{\cdot}6H_2O$, $FeCl_2{\cdot}4H_2O$ and deionized water. Then those two solutions mixed and heated to reaction temperature. Then the precipitant of $NH_3{\cdot}H_2O$ solution was added into the mixed solution under high speed stirring (1000 rpm) for 25 min until pH 5 value. After that, sodium oleate containing approximately 22 wt % Fe_3O_4 was introduced in deionized water with an ultrasonic mixer. Then this solution was added into the reaction solution which was adjusted to pH 5 under stirring with 1000 rpm at room temperature. At the end 40-min crystallization at 125 ^{0}C was carried out, followed by washing and freeze-drying. The chemical coprecipitation can be expressed as [6];

$$Fe^{2+} + 2Fe^{3+} + 8OH^- = Fe_3O_4 + 4H_2O \quad (1)$$

By using Eq.1, the pH value of the reaction was chosen to be 10 and the reaction temperature was 60^{0}C when Fe^{2+} precipitated completely. We used $NH_3{\cdot}H_2O$ as the precipitant to avoid rapid nucleus growth. The Fe^{2+}/Fe^{3+} ratio was chosen between 0.6-0.9 and concentration of ferric salt solution was set at 0.16 mol/L as constant. The drying process was performed in vacuum oven to avoid oxidizing of Fe^{2+}. For structural analysis, X-ray Difractometer (XRD), a Rigaku Dmax2 diffractometer with Cu Kα radiation (λ = 1.5406 Å) in the 2θ range from 20° to 80° with scanning rate of 0.2^o/s was used. Magnetic properties were carried out using Vibrating Sample Magnetometer (Quantum design VSM-9T) between ±9 T. Fourier transformed Infrared Spectrometer was used to identify the bond structure.

3.Result and Discussion

The results obtained and parameters used are summarized in Table 1. According to our experimental set up the best pH value was found to be 10 and during the experiments pH was set to constant value 10. However we know that the pH is an important factor for the chemical preparation of materials and the effect of pH on the materials is in progress however there exist valuable results in the literature explain the effects of pH for the nanomagnetic particles [12]. Figure 1 shows the XRD patterns of the Fe_3O_4 samples A1 and A4 with Fe^{2+}/Fe^{3+} ratio 0.6 and 0.9 respectively. As seen in Figures 1a and b, no impurities or secondary phase peaks were found in samples. According to calculations peaks were indexed as cubic Fe_3O_4 for all samples. However, intensity of the peaks is slight degreased and widened when average particle size decreased as expected.

Table 1. Summary of parameters used and results obtained.

Sample no	pH	Fe^{2+}/Fe^{3+}	Ferric salt Con. (mol/L)	Reaction Temp. (^{0}C)	Cryst. Temp. (^{0}C)	Average Particle size (nm)	Mag. (emu/g)	Room temp. Res. (ohm)
A1	10	0.6	0.16	60	75	6.9	3.1	1.0
A2	10	0.7	0.16	60	75	8.6	6.8	1.0
A3	10	0.8	0.16	60	75	10.8	11.4	1.2
A4	10	0.9	0.16	60	75	14.2	15.7	1.3

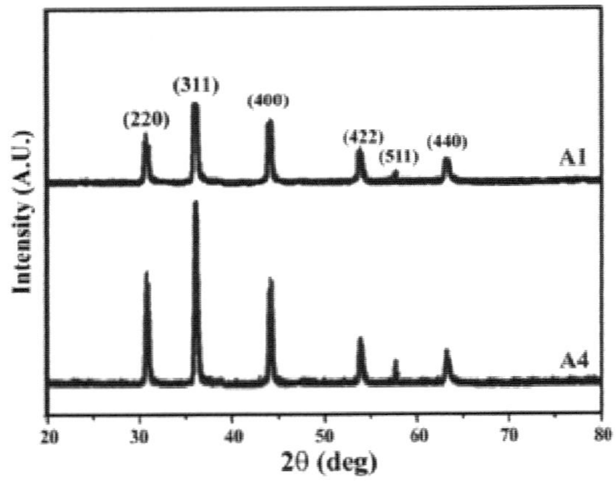

Figure 1. XRD results of a) sample A1 with particle size 6.9 nm and b) A4 14.2 nm.

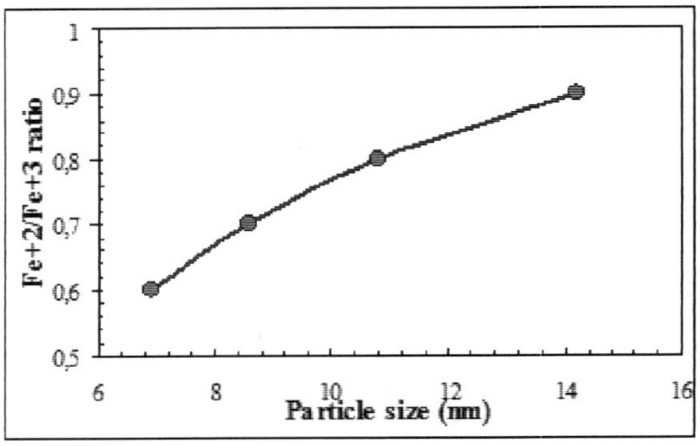

Figure 2. Particle size versus Fe^{2+}/Fe^{3+} ratio of the samples prepared.

The grain size of the samples prepared was determined by using Scherer equation [12-14];

$$d = \frac{0,89\lambda}{B\cos\theta} \qquad (2)$$

where, λ is the X-ray wavelength ($Cu_{K\alpha}$=1.5406Å), θ is the peak position and *B* is the peak full width at half-maximum intensity (*FWHM*). According to calculations with Eq.1 the size of the particles were found between 6.9 and 14.2 nm, Table 1. The results obtained showed that Fe^{2+}/Fe^{3+} ratio is playing a crucial role for the particle size. After third repeat of experiments we noticed that when Fe^{2+}/Fe^{3+} ratio increase the average particle size increases, Table 1 and Figure 2. As seen in Figure 2 when Fe^{2+}/Fe^{3+} ratio increase 50% the particle size increases approximately one fold. According to applications, for the biomedical use of these magnetic powders, needs to have particle size of less than 40 nm. Thought result obtained here in this work is reasonable at this stage. However, the effect of ferric salt concentration and reaction temperatures on the particle size and the magnetic properties is under progress.

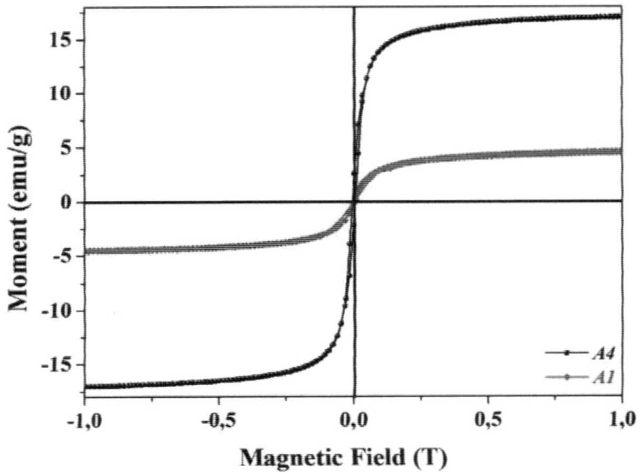

Figure 3. M-H curves of the sample A1 and A4 at room temperature.

Figure 3 shows the magnetization versus magnetic field graphs of sample A1 and A4. The hysteresis loops shows superparamagnetic nature. For both cases low saturation magnetization obtained and they showed no remanence, which is the typical property of Fe_3O_4 metallic samples. The saturation points were obtained at 3.1 and 15.7 emu/gr for A1 and A4 samples respectively indicates particle size effect for the magnetic properties of the samples prepared. Samples with small particle size showed low magnetization than the large particle samples. Figure 4.

Figure 5 shows the FT-IR graph of sample A4. All samples were showed similar trend and well known Fe-O vibration modes, 587, 1384 and 800 cm^{-1} of Fe_3O_4 were obtained. Whereas the peaks obtained at 1633 and 3408 cm^{-1} are attributed to the stretching vibrations of –OH and OH– absorbed Fe_3O_4 nano particles as obtained previously by other research groups [9, 12]. We did not obtain any other peaks between 1650 and 3400 cm^{-1} as obtained previously and assigned to the –CH, –C H_2 and –CH_3 [9,15].

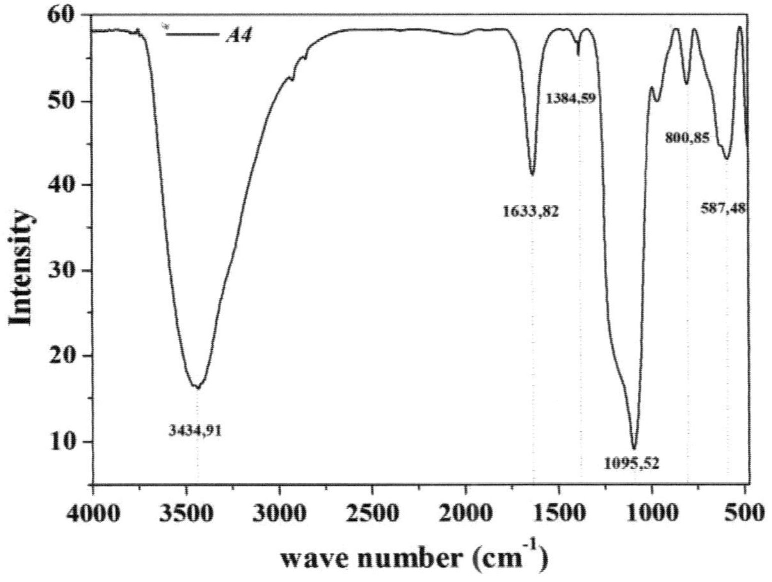

Figure 4. FT-IR graph of sample A4 at room temperature as an example.

4. Conclusion

We have successfully prepared high purity Fe_3O_4 nano particles using chemical coprecipitation method with particle size between 6.9 and 14.2 nm for biomedical applications. We found that particle size effective on the magnetic properties of the Fe_3O_4 nano material and this properties will be very important for the medical processes particularly for the cancer treatment of the defective cells. It was considered, according to magnetic properties of the prepared samples, the heat capacity of the small particles will be suitable for the heat absorbtion even during the small magnetic field applications than the larger particles. We did not get any different result on the room temperature resistance which is a positive effect for applications. We realized that chemical coprecipitation method is a suitable method for spherical and subnano Fe_3O_4 particles.

References

[1] Chieh J J *et al* 2015 *Applied Physics Letters* **106** 073703
[2] Oghabian M A *et al* 2010 *Nanomed.:Nanotechnol. Biol. Med.* **6** 496–499
[3] Müller S 2009 *Nanomed.:Nanotechnol. Biol. Med.* **5** 387–393
[4] Kim J S *et al* 2006 *Toxicol. Sci.* **89** 338–347
[5] Mody V V *et al* 2014 *Appl. Nanosci.* **4** 385–392
[6] Liu Y *et al* 2008 *Applied Surface Science* **255** 2020-2025.
[7] Zhi J *et al* 2006 *Reactive andFunctionalPolymers* **66** 1552-1558
[8] Liu Y *et al* 2009 *Applied Surface Science* **255** 7999-8002
[9] Meng H *et al* 2013 *Applied Surface Science* **280** 679– 685
[10] Hong R *et al* 2007 *China Particuology* **5** 186–191
[11] Kim D K *et al* 2001 *Journal of Magnetism and Magnetic Materials* **225** 30-36
[12] Sun J *et al* 2007 *Journal of Biomedical Materials Research Part A* **2** 333-341
[13] Ahmad S *et al* 2009 *J Inorg Organomet Polym* **19** 355–360
[14] Ghandoor H E *et al* 2012 *Int. J. Electrochem. Sci.,***7** 5734-5745
[15] Jordan A *et al* 1996 *International Journal of Hyperthermia* **12** 705–722

Solid state synthesis and characterization of bulk $FeTe_{0.5}Se_{0.5}$ superconductors

K. Onar[1] and M. E. Yakinci[1,2]

[1]İnönü Üniversitesi, Fen Edebiyat Fakültesi, Fizik Bölümü, 44280-Malatya, TURKEY
[2]İnönü Üniversitesi, Mühendislik Fakültesi, Biyomedikal Mühendisliği Bölümü, 44280-Malatya, TURKEY

E-mail: kubra.onar@inonu.edu.tr

Abstract. $FeTe_{0.5}Se_{0.5}$ polycrystalline superconductor samples were synthesized by solid-state reaction method at different heating temperatures. The morphological and structural characterization of $FeTe_{0.5}Se_{0.5}$ samples were carried out by X-rays Diffraction, Scanning Electron Microscope and Energy Dispersive X-ray Spectroscopy. The electrical, magnetic and thermal transport properties were investigated up to 8 T by using physical property measurement system. The results reveal that the sensitivity of electrical and magnetic properties strongly depends on the heat treatment cycles. The upper critical field, $H_{c2}(0)$, was determined with the magnetic field parallel to the sample surface. It gives a maximum value of 36.3 T. The lower critical field, $H_{c1}(T)$, was obtained as 210, 140 and 70 Oe at 5, 8 and 12 K, respectively. The coherence length, ξ, at the zero field, was calculated to be 1.94 nm and suggested a transparent intergrain boundaries peculiarity. The $\mu_0 H_{c2}(0)/k_B T_c$ rate shows higher value (3.36 T/K) than the Pauli limit (1.84 T/K) which suggests unconventional nature of superconductivity for the polycrystalline $FeTe_{0.5}Se_{0.5}$ superconducting samples.

1. Introduction

Superconductivity and magnetism are two different concepts in physics and they are not considered together, because superconductivity is disappeared by the magnetism, until late 1990. When Ru based materials were first discovered in 1997, the scientists were surprised due to existence of both magnetism and superconductivity together in the sample [1]. Then in 2008, discovery of $LaFeAsO_{1-x}F_x$ with superconducting transition temperature T_c up to 26 K attracts a big attention due to existence of magnetic Fe element and superconductivity in the material [2]. After that different phases of iron based superconductors were discovered, $Ba_{1-x}K_xFe_2As$, the "122-phase" with T_c of ~38 K [3], $Li_{1-x}FeAs$, so-called "111-phase" with T_c=18 K [4] and FeSe, so-called "11-phase" with T_c=8 K [5]. The superconducting transition temperature has been increased up to ~55 K by replacing or substituting/doping of La with other rare earth elements such as, Ce, Sm, Nd, Pr, Gd, Eu and Tm. The obtained phase is then called "1111-phase" [6, 7]. The FeSe (11-phase) has the most simple crystal structure, T_c's up to 37 K under high pressure, and most importantly it has not contain toxic materials such as, As and it is easy to obtain in single crystalline form [8-13].

In general, iron-based superconductors have layered crystal structure [14]. Fe-As layers, which are negatively charged, and Ln-O layers (Ln: lanthanides), positively charged, are simply arranged along the c-axis. The Ln-O layers are insulating nature and the Fe-As layers are responsible from the conducting mechanism [14,15]. The FeSe phase, called as "11-phase" with T_c=8 K, which is a derivative of FeAs family, is composed of tetrahedral stacks of Fe_2Se_2 layers which have edge sharing peculiarity and behave as conducting layers similar to B_2 layers in MgB_2 metallic superconducting alloys [7,16]. Recently, alkaline metals such as, K, Rb, Cs, Li and Na have been added successfully into FeSe materials. The physical properties of obtained materials show that all additives behave as a charge carriers and $(Fe_2Se_2)^{\delta-}$ layers are responsible at the conductivity [17-20]. Mizuguchi et. al. obtained FeTe material which has tetragonal crystal structure that is similar to FeSe superconductors [21] and it has a magnetic or a structural phase transition at around 70 K [22, 23]. However, FeTe does not show superconductivity. The other investigation indicates that Se substitution suppressed and causing superconductivity [24-27]. The FeTe materials can be prepared either in polycrystalline or in single crystal form by using different methods [28-30]. In the application side of FeSe and FeTeSe materials, various forms such as, thin/thick films [31, 32] and wires/tapes [33-36] can be prepared and they have capabilities to meet today's technological needs.

In this work, polycrystalline $FeTe_{0.5}Se_{0.5}$ samples have been prepared by using conventional solid state reaction method. To understand better the physical and magnetic properties, we have focused on the estimations of the upper critical field (H_{c2}), lower critical field (H_{c1}), activation energy (E_a) and the coherence length, ξ, as well as the structural and electrical transport properties.

2. Experimental Details

Polycrystalline $FeTe_{0.5}Se_{0.5}$ material was prepared using conventional solid-state reaction method. High purity powders of the Iron (99.999% Alfa Aesar), Tellurium (99.999% Alfa Aesar) and Selenium (99.999% Alfa Aesar) were used as raw materials. Appropriate amount of Fe/Te/Se:1/0.5/0.5 was ground by agate mortar to obtain a homogenized mixture and then transferred into quartz tubes and sealed.

Then the quartz tubes were placed inside a tube furnace and heated up with a rate of 2°C/min at 600, 650 and 700°C (so called samples FTS-1, FTS-2 and FTS-3 respectively) for 24 h. After slow cooling to room temperature the first heating cycle was repeated again with an intermediate grinding in a glove box under Ar atmosphere. In the third and final heating stage, sample was re-grounded and pressed isostatically in the glove box at 65 MPa and annealed at fixed 400°C for 24 h with 2°C/min heating and cooling ramps in the evacuated quartz tubes, Table1. After the heating cycles, structural properties of the $FeTe_{0.5}Se_{0.5}$ samples obtained were examined by X-Rays Powder diffractometer (XRD). A Rigaku RadB DMaxII difractometer and Jade 6+ refinement program was used to analyze phase formation. The microstructural formations of the samples prepared were investigated using Scanning Electron Microscope (SEM) (Leo EVO 40VPX)

Table 1. Heat treatment cycles of polycrystalline $FeTe_{0.5}Se_{0.5}$ samples.

	Heating stages					
Sample	**First heat treatment**		**Second heat treatment**		**Third heat treatment**	
	Temp. (°C)	*Time (h)*	*Temp. (°C)*	*Time (h)*	*Temp. (°C)*	*Time (h)*
FTS-1	600	24	600	24	400	24
FTS-2	650	24	650	24	400	24
FTS-3	700	24	700	24	400	24

and Energy Dispersive X-rays spectroscopy (EDX) (Bruker 125 eV). The electrical, thermal and magnetic properties were carried out using Quantum Design Physical Properties Measurement Systems, (PPMS-9T) with applied fields up to 8 Tesla. To determine the upper critical field $H_{c2}(0)$, the *R-T* curves, measured under applied fields up to 8 T and Werthamer-Helfand-Hohenberg (WHH) formula, were used. The superconducting coherence length was estimated using Ginzburg Landau (GL) equation. The *I-V* characteristics of the samples were determined using 1 μV criteria. The activation energy of the samples was calculated using the Arrhenius equation.

3. Results and discussions

3.1. X-ray analysis (XRD)

Figure-1 shows the XRD patterns of the polycrystalline $FeTe_{0.5}Se_{0.5}$ samples prepared by solid state reaction method. Heat treatment cycles of the samples prepared are summarized in Table 1. A slight difference between the x-ray patterns of samples heat treated at 600 to 700 °C was obtained. As seen in Figure-1, all of the polycrystalline $FeTe_{0.5}Se_{0.5}$ samples prepared contain small amount of impurity phases. These phases are mainly antiferromagnetic hexagonal $FeTe_{0.5}Se_{0.5}$ phase (marked with H) and unreacted Fe element (marked as asterisks). We did not detect common impurity phases such as, $FeSe_2$, Fe_7Se_8 and δ-FeSe and also oxide peaks as obtained in previous works by other groups [27, 37, 38]. The main peaks are indexed as tetragonal $FeTe_{0.5}Se_{0.5}$ superconducting phase with a space group of P4/nmm, Lattice parameters and unit cell volume of the samples prepared were calculated to be a=3.7885(5) Å, c=5.9453(5) Å and 85.03 $Å^3$ respectively. After the optimization of heating cycles the intensity of the impurity phases was decreased but, single phase samples were not obtained, Figure 1. This indicates that the structural formation of the polycrystalline $FeTe_{0.5}Se_{0.5}$ samples was not largely depending on the heat treatment cycles. Similar results were obtained previously by different groups and can be associated with the large melting temperature difference between the Fe, Se and Te materials [36, 39].

3.2. SEM and EDX analysis

Figure 2a and 2b show the surface micrographs of the sample FTS-2 in different magnifications as an example. Strongly connected but randomly oriented granular formation with an average grain size of 2 μm, which is the typical growth characteristics of polycrystalline FeTeSe material prepared with the solid state reaction method, were obtained.

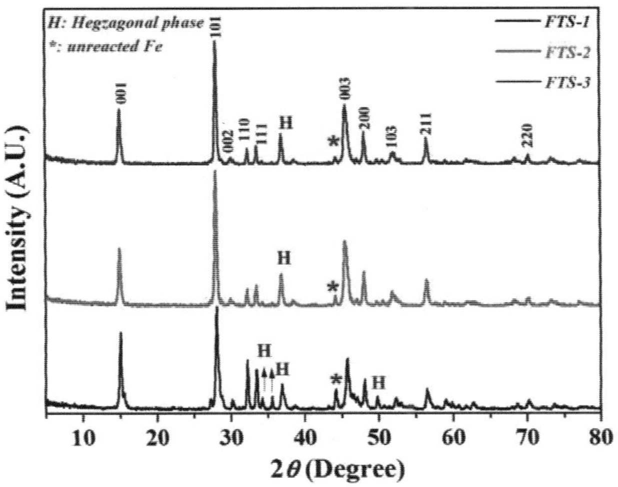

Figure 1. XRD graphs of the polycrystalline $FeTe_{0.5}Se_{0.5}$ samples heated at different temperatures.

Journal of Physics: Conference Series **667** (2016) 012006 doi:10.1088/1742-6596/667/1/012006

Microstructural formation, show that there is no effect of thermal treatment on the microstructural evolution as obtained on the XRD results. We also noticed that the density of the samples was not high comparing to other metallic structures or single crystalline form of same material. According to EDX results calculated ratio of Fe:Te:Se was found as 49.6:25.2:25.2 for sample FTS-2 and FTS-3. These data are very close to the 1:0.5:0.5 starting ratio and indicating that the good reaction between the Fe, Te and Se in the quartz tubes. The sample FTS-1 showed slightly Fe rich and Te deficient phase formation and the deviation from the actual composition is about 3%, however, which has no effect on the microstructural formation of the samples. The EDX dot mapping for sample FTS-3 are given in Fig. 2c as an example and homogenious $FeTe_{0.5}Se_{0.5}$ phase distribution was obtained.

3.3 Transport Properties

Figure 3a shows temperature dependence of resistivity for the polycrystalline $FeTe_{0.5}Se_{0.5}$ samples at zero field. The region around the $T_c^{(onset)}$ is magnified and shown in the inset of Figure 3a. In generate the results obtained are summarized in Table 2. According to Figure 3a, all samples have a metallic behavior down to their $T_c^{(onset)}$ temperature. Samples FTS-2 and FTS-3 are showed a sharp resistivity drop after $T_c^{(onset)}$ showing that the purity and the quality of the samples. A double transition was observed for the sample FTS-1, indicating a multiphase nature as obtained in the XRD and EDX results in the previous sections. The first transition, obtained at around 14 K, is believed to be due to the tetragonal FeTeSe superconducting phase, and the second transition, obtained at around 10.3 K, indicates the existence of the impurities, mainly hexagonal FeTeSe phase and unreacted Fe in micro-level. This indicates insufficient ionic diffusion process due to low temperature heating of this sample comparing to the FTS-2 and FTS-3. Although anomalies such as, structural or magnetic phase transformation between 50 K and $T_c^{(onset)}$ for all samples was observed. In order to determine superconducting transition temperature of the samples precisely *dρ(T)/dT* values were calculated.

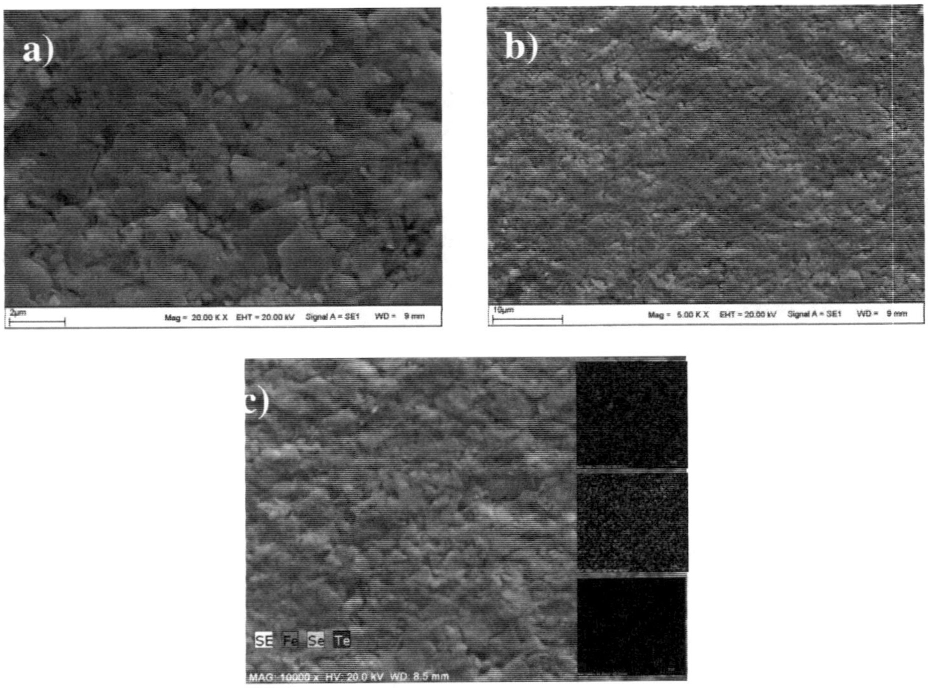

Figure 2. SEM micrographs of the polycrystalline $FeTe_{0.5}Se_{0.5}$ samples **a)** high magnification image **b)** Low magnification image **c)** EDS mapping image.

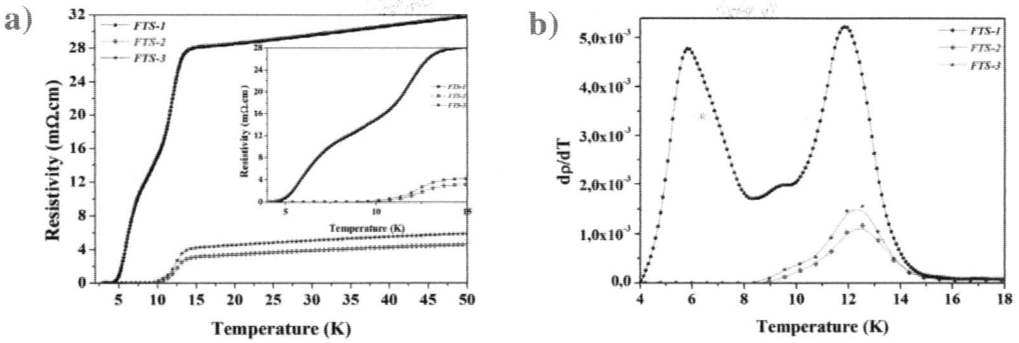

Figure 3. a) Temperature dependence of resistivity for the polycrystalline $FeTe_{0.5}Se_{0.5}$ samples, **b)** derivative of resistivity dρ/dT as a function of temperature.

Figure 3b shows the $d\rho(T)/dT$ versus $T(K)$ graphs of samples and the results are given in Table-2. Figure 3b indiçates that, sample FTS-2 has much sharper peak than those of FTS-3 and FTS-1, suggesting that the heat treatment condition of the sample is the optimum. This peculiarity is also confirmed in the graphs of $\rho(T)$-$T(K)$, in Figure 3a, as high $T_c^{(onset)}$ and T_{zero} values with a sharp transition. However, sample FTS-1 was showed a double peak, means that the double transition as obtained in the graphs of $\rho(T)$-$T(K)$, in Figure 3a, suggest that the multiphase nature as mentioned before. Figure 4a shows the resistivity transition at magnetic field up to 8 T with applied field, H, parallel to the sample surface for sample FTS-2 as an example. The other samples were showed similar trends. In general, the resistive transition was shifted to a lower temperatures and superconducting transition is slightly broadened by increasing the magnetic field as expected. Because, an increase on the magnetic field, attenuate the electron pairs by enforcing to bring the opposite spins in the same direction. This causes a reduction on the electron pairs population, reduces the T_c and delays the transition to the superconducting state. To determine the upper critical field $H_{c2}(0)$, we used 10% $T_c^{(offset)}$, 50% $T_c^{(midpoint)}$ and 90% $T_c^{(onset)}$ criterion and Werthamer-Helfand-Hohenberg (WHH) formula. The WHH formula is given as [39];

$$H_{c2}(0) = -0{,}693T_c\left(\frac{dH_{c2}(T)}{dT}\right)_{T=T_c} \quad (1)$$

By using Eq-1, the zero temperature value of upper critical field, $H_{c2}(0)$, is calculated to be 87.5, 69.9 and 36.3 T for 10% $T_c^{(offset)}$, 50% $T_c^{(midpoint)}$ and 90% $T_c^{(onset)}$ respectively, Figure 4b. These values are slightly higher than the literature value [40]. The theoretical value of $H_{c2}(T)$ was also calculated using an equation which is developed by Wen et.al. [41];

Table 2. Superconducting properties of polycrystalline $FeTe_{0.5}Se_{0.5}$ samples prepared.

Sample No	$T_c^{(onset)}$ (K)	$T_c^{(zero)}$ (K)	ΔT (K)	$H_{c2}(0)$ (T) %10-%50-%90	ξ(0) (nm) %10-%50-%90	$E_a(H)$ (0T)	J_c^{trans} (Acm^{-2}) at 4.2 K (0T)
FTS-1	14.3/10.3	5.2	9.1	66.6 - 50.6 - 23.6	2.22 - 2.55 - 3.73	2.610^{-13}	0.28
FTS-2	15.6	9.5	6.1	87.5 - 69.9 - 36.3	1.94 - 2.17 - 3.01	1.310^{-12}	6.06
FTS-3	14.9	8.3	6.6	84.1 - 57.7 - 32.8	1.98 - 2.38 -3.17	9.410^{-13}	2.09

$$H_{c2}(T) = H_{c2}(0)\frac{\left[1-\left(T/T_c(0)\right)^2\right]}{\left[1+\left(T/T_c(0)\right)^2\right]} \quad (2)$$

For sample FTS-2, the graph of calculated values are given in insert (lower left) of Figure 4b, as an example. At 0 K, calculated values were found to be 84.04 T, 57.60 T and 32.59 T for 10% $T_c^{(offset)}$, 50% $T_c^{(midpoint)}$ and for 90%$T_c^{(onset)}$ respectively. These figures are found close to the results calculated by using the WHH equation. The superconducting coherence length, (ξ), along the *c*-axis, was estimated using Ginzburg Landau (GL) equation. The GL equation is given by;

$$H_{c2}(0) = \frac{\phi_0}{2\pi\xi^2(0)} \quad (3)$$

$$\xi(0) = \left(\frac{\phi_0}{2\pi H_{c2}(0)}\right)^{1/2} \quad (4)$$

here, $\Phi_0 = 2.07 \times 10^{-15}$ Wb is the flux quantum [42, 43]. The calculated values of ξ(0) for 10%$T_c^{(offset)}$ criterion are obtained to be 1.94, 1.98 and 2.22 nm for sample FTS-2, FTS-3 and FTS-1 respectively, Table-2. These results were found close to the previously calculated values by other research groups and also comparable to the single crystal samples [43]. We also calculated the $\mu_0 H_{c2}(0)/k_B T_c$ value, which gives valuable information about the nature of the materials by using $H_{c2}(0)$ obtained from the WHH theory. The $\mu_0 H_{c2}(0)/k_B T_c$ results obtained showed comparably higher value (3.36 T/K) than the Pauli limit (1.84 T/K). This suggested unconventional nature of superconductivity in the sample as obtained in the FeTeSe single crystal family and also FeSe polycrystalline materials [30,36,44].

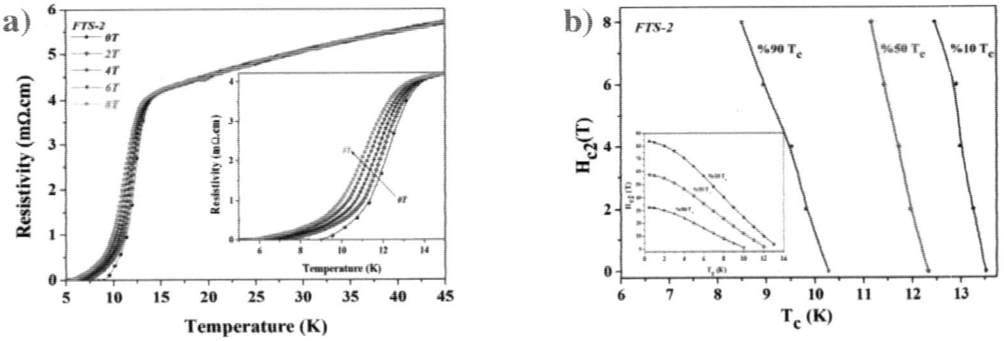

Figure 4. a) Field dependence of the resistivity of sample FTS-3 **b)** the upper critical field, H_{c2}, as a function of critical temperature. The insert shows calculated theoretical values.

3.4. The Activation Energy Calculations

The activation energy, E_a, is an important parameter for the description of the flux dynamics in the mixed state and revealing the height of the energy barriers which controls the vortex motion in the vortex liquid. In the thermally activated flux flow (TAFF) region, the thermally activated resistivity is defined as [45],

$$\rho = \rho_0 exp[-\frac{E_a(J,H,T)}{k_B T}] \quad (5)$$

where ρ_0 is the pre-exponent factor, k_B the Boltzmann constant and E_a the activation energy for the flux motion that generally depends on the temperature, T, the current density, J, and the applied magnetic fields, H. At constant current, the functional dependence of E_a (H,T), is believed to give information about the dissipation mechanism and can be calculated from the slope of $\ln(\rho/\rho_0)$ versus $1/T$ graphs [45]. Figure 5 shows the E_a versus $H(T)$ graphs of the samples FTS-1, FTS-2 and FTS-3. In general, the variation of E_a is explained as an increase of flux trapping due to increased magnetic field. It is also well known that the decrease of the activation energy promotes the thermally activated dissipation due to the vortex motion in a single-crystal sample [45]. In contrast, for polycrystalline superconductors, dissipation begins in intergrain boundaries instead of sample surface or inside the sample. Thus, low value of E_a can be explained with the transparent intergrain boundaries.

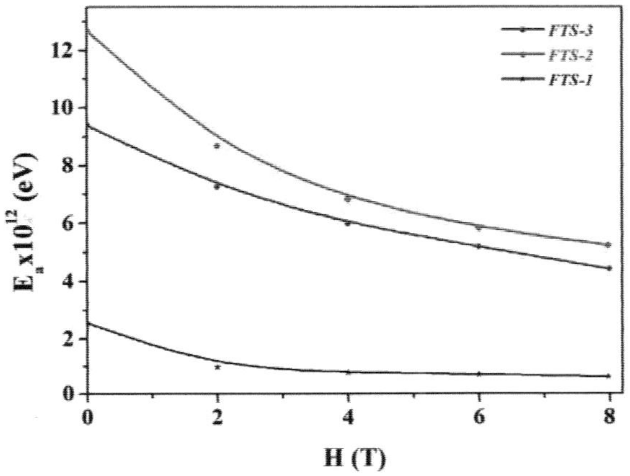

Figure 5. Calculated activation energy of the samples as a function of magnetic field.

3.5. Magnetic Properties

Figure 6a shows the temperature dependence of the magnetization (*M-T*) measurements under 10 Oe for samples prepared with different heating cycles. Above T_c, sample FTS-1 showed Pauli paramagnetic type behavior up to ~125 K and then remain unchanged down to ~10 K, then dropped to diamagnetic phase just after T_c. This indicates the role of impurities ie: unreacted Fe and hexagonal non-superconducting phase of FeTeSe material on the magnetic properties. An anomaly, which is obtained at around 125 K, is believed to be related with the structural transformation. However, sample FTS-2 and FTS-3 showed negative ferromagnetic sign down to T_c and then sharp drop was obtained just after T_c, Figure 6a.

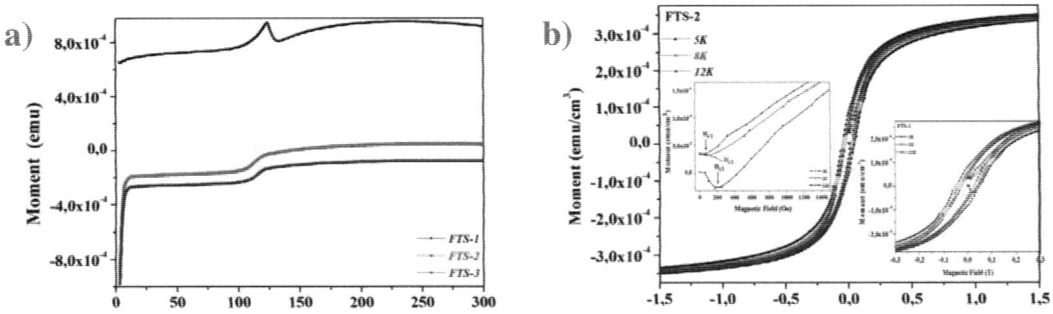

Figure 6. a) Temperature dependence of Magnetization and **b)** M-H curves at different temperatures for sample FTS-2. The first quadrants of the M-H curves are given in the left insert.

The controversy between the sample FTS-1 and both FTS-2 and FTS-3 suggests the role of different spin-spin and/or spin-orbital interaction of different phases, even under small magnetic fields (10 Oe) during the measurements. Magnetic hysteresis curves of sample FTS-2 as an example at 5, 8 and 12 K are given in figure 6b. It can be seen that polycrystalline FTS-2 sample has mainly a ferromagnetic background due to ferromagnetic nature of Fe, which yields more dominant ferromagnetic signal than the diamagnetic signal. In general, this indicates the absence of bulk superconductivity in the samples, as obtained by the other research groups [35-41]. The lower critical field (H_{c1}) is an important superconducting parameter where the magnetic field starts to penetrate in to the sample and accordingly vortexes consist of on the sample surface as expected. The first quadrant of the M-H curves at 5, 8 and 12 K were measured to determine the lower critical field. The results of sample FTS-2 are given in Figure 6b (the left insert) as an example. The lower critical field values of sample FTS-2 were calculated as 210, 140 and 70 Oe for 5, 8 and 12 K, respectively. These values were found to be lower than the HT_c materials but are comparable to low T_c conventional superconductors and FeSe based materials.

3.6. I-V Measurements

The I–V characteristics of the samples were measured by the four probes technique with the current flow along the sample surface. The measurements were carried out at 4.2 K with magnetic fields ranging from 50 Oe to 1T. I–V curves of the sample FTS-2 at 4.2 K, under different magnetic fields. The results are given in Figure 7 and summarized in Table 2. The I–V characteristic of the samples appear as flux creep behavior as obtained for the HT_c BSCCO material [45]. The highest J_c^{trans} value was calculated to be 60.6 A/cm^2 at zero field and 4.2 K and 19.7 A/cm^2 at 1 T and 4.2 K. When the external field increases, the critical current density, J_c^{trans}, decreases in the flux creep regime. However, the subsidence is found to be very slow comparing with the HT_c and also other conventional superconductors and suggested strong grain connectivity properties in this material. In addition we believed that The small amount of impurity phases, which provides pinning centers on the surface, play a crucial role on the J_c^{trans} limit particularly for sample FTS-2 and FTS-3.

3.7. Thermal properties

The Seebeck coefficient, S, of the polycrystalline $FeTe_{0.5}Se_{0.5}$ samples are given as a function of temperature in Figure 8a. The Seebeck coefficient, S, of samples FTS-2 and FTS-3 was started from negative value (approximately -2μV/K) at high temperatures and then slightly increases between 250 and 200 K. After the certain temperature it drops to ~25K and then increases again. At low temperature region, the sample FTS-1 shows similar trend like FTS-2 and FTS-3 samples. But a positive S value is obtained between 180 and 290 K and then it turns to negative value at 300 K. This anomaly can be explained using drag mechanism which is related with the exchange of momentum between charge carriers and magnetic fluctuations of particularly ferromagnetic phase that is dominant all over the sample. The temperature dependence of thermal conductivity, κ, are shown in Figure 8b. The κ value of the samples show a variation temperature at room temperature. For the sample FTS-2, the optimally treated sample, indicates comparably high κ value at room temperature. It decreases slowly from the room temperature to 75 K and then rapid reduction is observed to the minimum value. The other sample show similar peculiarity. In general, the obtained results show a classical property of metals and alloys and suggesting two important points. Firstly; at high temperature region, thermal conductivity dominates by phonons because electron contribution evaluates by the Wiedermann-Franz law being negligible as suggested by other research groups [46]. Secondly, at low temperature region the main scattering mechanism of phonons was charge carriers and structural defects such as, a large Fe vacancies and relatively large c-axis disorders.

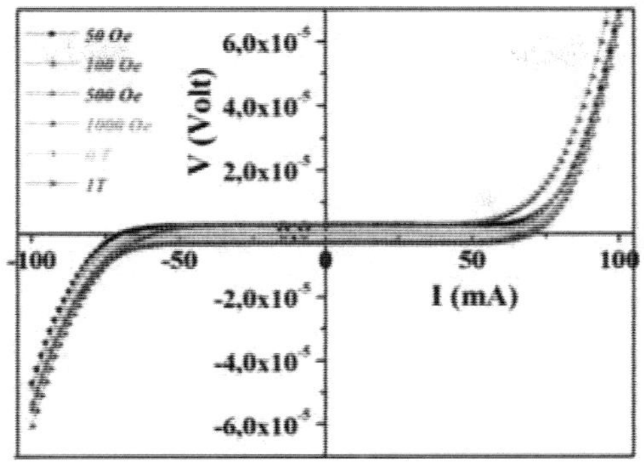

Figure 7. The I-V characteristics of the sample FTS-2 at different magnetic fields.

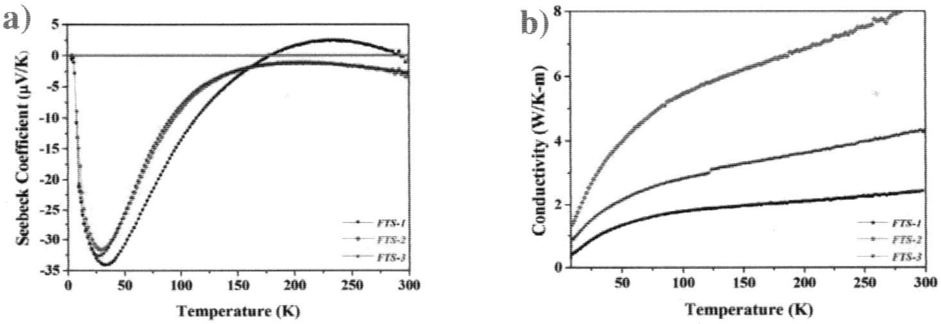

Figure 8.a) Temperature dependences of the Seebeck coefficient, (*S*), and **b)** Temperature dependences of the thermal conductivity, (κ), of the samples prepared.

4. Conclusion

The Polycrystalline $FeTe_{0.5}Se_{0.5}$ superconductors have been prepared by conventional solid-state reaction method in the evacuated quartz tube. Structural, physical, magnetic and transport properties were investigated. The results show that FeTeSe polycrystalline samples are very sensitive to the magnetic field and this intricacy can be related with insufficiency of bulk superconductivity in the samples where different magnetic behaviors can be occurred. For optimally treated sample (FTS-2) the upper critical field, $H_{c2}(0)$, was obtained to be 87.5 T, at the zero field coherence length, ξ, value was calculated to be 1.94 nm and the lower critical field, $H_{c1}(T)$, is obtained to be 210, 140 and 70 Oe at 5, 8 and 12 K, respectively.

The $\mu_0 H_{c2}(0)/k_B T_c$ value, which gives more information about the nature of the superconductor materials showed higher value (3.36 T/K) than the Pauli limit (1.84 T/K). In overall, the results indicate the unconventional nature of superconductivity in the polycrystalline $FeTe_{0.5}Se_{0.5}$ superconductor samples. The J_c value of samples is still low but it is convenient for bulk applications such as, large magnets or flywheels.

Acknowledgment

This work is partially supported by the İnönü Üniversitesi Bilimsel Araştırma Projeleri Koordinasyon Birimi Project No: 2012/41 GÜD.

References

[1] Felner I *et al* 1997 *Phys. Rev. B* **55** R3374
[2] KamiharaY *et al* 2008 *J. Am. Chem. Soc.***130** 3296
[3] Hsu F C 2008 *Proc. Nat. Acad. Sci.USA* **105** 14262
[4] ChenX H *et al* 2008 Nature **453** 761
[5] Chen G F *et al* 2008 *Phys. Rev. Lett.***100** 247002
[6] Rotter M *et al* 2008 *Phys. Rev. Lett.* **101** 107006
[7] Wang X C *et al* 2008 *Solid State Commun* **148** 538
[8] Ishida K *et al* 2009 *J. Phys. Soc. Japan* **78** 062001
[9] Yeh K W *et al* 2009 *Cryst. Growth Des.* **9** 4847
[10] Yadav C S *et al* 2009 *J Phys* **11** 103046
[11] Kida T *et al* 2009 *J. Phys. Soc. Japan* **78** 113701
[12] Braithwate D *et al* 2010 *J. Phys. Soc. Japan* **79** 053703
[13] Taen T *et al* 2010 *Physica C* **470** 391
[14] Deguchi K *et al* 2012 *Adv. Mater.* **13** 054303
[15] Aswathy P M *et al* 2010 *Supercond. Sci. Technol.* **23** 073001
[16] Aksan M A *et al* 2004 *Physica C* **408** 132-133
[17] Wang A F *et al* 2011 *Phys. Rev.B* **83** 060512(R)
[18] Li C H *et al* 2011 *Phys. Rev. B* **83** 184521
[19] Ying J J *et al* 2010 *Phys. Rev. B* **83** 212502
[20] Guo J *et al* 2010 *Phys. Rev. B* **82** 180520(R)
[21] Mizuguchi Y *et al* 2009 *Physica C* **469** 1027–1029
[22] Li S *et al* 2009 *Phys. Rev. B* **79** 054503
[23] Bao W *et al* 2009 *Phys. Rev. Lett.* **102** 247001
[24] Yeh K W *et al* 2008 *Europhys. Lett.* **84** 37002
[25] Fang M H *et al* 2008 *Phys. Rev. B* **78** 224503
[26] Mizuguchi Y *et al* 2009 *Phys. Soc. Japan* **78** 074712
[27] Ding Q P *et al* 2011 *Supercond. Sci. Technol.* **24** 075025
[28] Williams A J *et al* 2009 *Phys.,Condens. Matter* **21** 305701
[29] Guler N K *et al* 2014 *J Supercond Nov Magn* **27** 2691-2697
[30] Zhang S B *et al* 2009 *Supercond. Sci. Technol.* **22** 015020
[31] Wang M J *et al* 2009 *Phys. Rev. Lett.* **103** 117002
[32] HanY *et al* 2009 *Phys.,Condens. Matter* **21** 235702
[33] Togano K *et al* 2012 *Solid State Communications* **152** 740–746
[34] Ma Y *et al* 2012 *Supercond. Sci. Technol.* **25** 113001
[35] Ozaki T *et al* 2011 *Supercond. Sci. Technol.* **24** 105002
[36] Mizuguchi Y *et al* 2011 *Supercond. Sci. Technol.* **24** 125003
[37] Wu M K *et al* 2013 *Sci.Technol. Adv. Matter.* **14** 014402
[38] Migita M *et al* 2011 *Physica C* **471** 916–918
[39] Onar K *et al* 2015 *J of Alloys and Compounds* **620** 210–216
[40] Zargar R A *et al* 2014 *J Supercond Nov Magn* **27** 897–901
[41] Fang L *et al* 2005 *Phys. Rev.B* **72** 014534
[42] Lei H C *et al* 2010 *Phys. Rev.B* **81** 094518
[43] Lei H C *et al* 2010 *Phys. Rev. B* **81** 184522
[44] Velasco Soto D *et al* 2013 *J. of Appl. Phys.* **113** 17E138
[45] Aksan M A *et al* 2007 *Solid Films* **515** 8022–8027
[46] Tropeano M *et al* 2010 *Supercond. Sci. Technol.* **23** 054001

Tilted vortex lattice in irradiate $Bi_2Sr_2CaCu_2O_{8+\delta}$ single crystals

J Mirkovic[1,2], I Kakeya[3], S Savel'ev[4], T Kashiwagi[2], B Markovic[5], K Kadowaki[2]

[1] Faculty of Science, University of Montenegro, G. Washington str. bb, Podgorica, and Centre for Ecotoxicological Research of Montenegro, 81000 Podgorica, Montenegro
[2] Division of Materials Science, Faculty of Pure and Applied Sciences, University of Tsukuba, Tennodai, Tsukuba, Ibaraki 305-8573, Japan
[3] Department of Electronic Science and Engineering, Kyoto University, Nishikyo-ku, Kyoto 615-8510, Japan
[4] Department of Physics, Loughborough University, Loughborough, LE11 3TU, UK
[5] Faculty of Electrical Engineering, University of Montenegro, G. Washington str. bb, and Centre for Ecotoxicological Research of Montenegro, 81000 Podgorica, Montenegro

E-mail: jovan.mirkovic@prona.org

Abstract. In order to enlighten the structure of vortex matter in irradiated layered $Bi_2Sr_2CaCu_2O_{8+\delta}$ single crystals, the interaction of Josephson vortices and pancake vortices in was investigated by means of the local ac-magnetic permeability measurements by using the miniature local coils, while vortex matter in pristine crystals was studied by in-plane resistivity measurements. The transition anomaly, separating the strong pinning phase and the weak pinning vortex phase was found by both techniques deep in the vortex solid phase solid near ab-plane, indicating crossover from the vortex chains + lattice phase to tilted vortex chains phase. While the columnar defects affect strongly the first-order vortex-lattice melting transition, the magnetic permeability anomaly, associated with the crossover from vortex chains + lattice phase to tilted lattice, is surprisingly still clear, deep in the vortex solid phase. However, the stronger columnar defects eventually affect the crossover anomaly that it disappears too.

1. Introduction

Vortex matter in layered superconductors, such as $Bi_2Sr_2CaCu_2O_{8+\delta}$ single crystals, differs from vortex lattices in moderately anisotropic superconductors, such as $YBa_2Cu_3O_{7-\delta}$. In tilted magnetic fields, $Bi_2Sr_2CaCu_2O_{8+\delta}$ samples have two types of vortices: the c-axis component H_c of the magnetic field produces (2D) pancake vortices (PVs) [1], while the in-plane field component H_{ab} generates Josephson vortices (JV) between the CuO_2 planes [2,3]. Interacting crossing PV and JV lattices have been studied theoretically [3]-[7] and recently were observed by using Lorentz microscopy [8,9], scanning Hall probe microscopy [10], and magneto-optical imaging [11]. There are two reasons of this surge of the interest to vortex matter and in particular crossing vortex lattice in layered superconductors. First, crossing vortex lattices exhibit non-trivial and fascinating dynamical behavior (see,e.g., [12,13]) which allows us to manipulate [14]-[18] one vortex lattice via another one (or in more general interacting nanoparticles), as well as non-standard thermodynamic properties [19-25]. For instance, in contrast to the Ginzburg–Landau (GL) approach, the first-order melting transition line (the boundary separating the vortex-solid and vortex-liquid phases) exhibits a 'step-wise' shape [19] (i.e., approximately linear decay interrupted by two flat-plateau) if plotted in the H_c–H_{ab} plane.

Namely, the *c*-axis component H_{trans} of magnetic field at the phase transition decays linearly when increasing H_{ab}, followed by a plateau. A similar stepwise shape of the phase boundary was observed [26] at a much lower temperature, where disorder plays an essential role and the transition is smooth, in contrast to first-order. Thus, this suggests that the shape of the transition line is controlled by general thermodynamic properties, rather than the precise statistical features of the vortex-solid-phase, including vortex-glass, Bragg-glass, or Bose-glass. The second reason of huge interest to vortex phase diagram in Bi-compounds is related to possible application of this material for superconducting THz electronics [27-38].

In this article we report recent measurements of the vortex-solid phase diagram in $Bi_2Sr_2CaCu_2O_{8+\delta}$ pristine and irradiate samples and observation of the phase boundary deep in the vortex solid phase which can be attributed to a vortex structure transition. This transition smears out for strongly irradiated samples. We discuss this transition based on the vortex phase diagram having both tilted and crossing vortex lattice structures.

2. Experiment

The two $Bi_2Sr_2CaCu_2O_{8+\delta}$ as-grown single crystals with T_c = 89 and 88.9 K were used . These samples had defects along *c*-axis introduced by heavy-ion irradiation ($_{127}I_{28+}$,650 Mev), with doses of B_Φ = 0.1 and 0.02 T, respectively. The vortex phases were probed via local ac mutual-inductance measurements by using a set of two miniature coils. The size of the coils was 0.35 mm, which was sufficiently smaller than the size of the samples (~3mm × 3mm) in order to diminish the surface barrier.

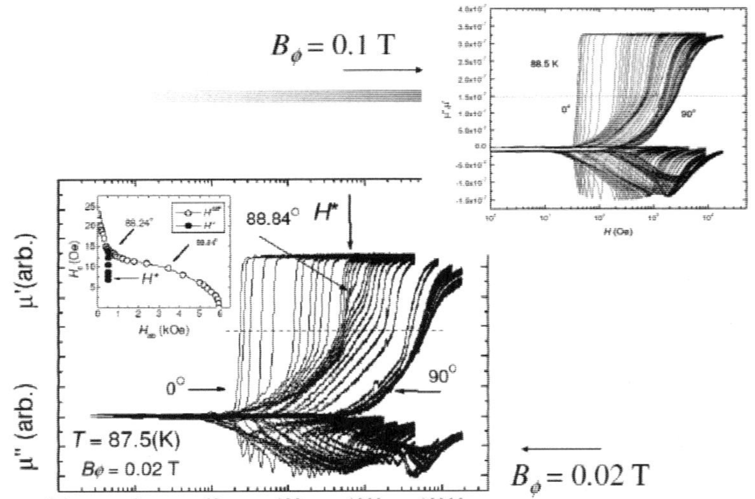

Figure 1. The magnetic field dependence of the real and imaginary parts of the ac-local magnetic permeability μ for irradiated $Bi_2Sr_2CaCu_2O_{8+\delta}$, with two doses B_Φ = 0.1 T (upper) and B_Φ = 0.02 T (lower) with corresponding phase diagram (left).

The dc magnetic field, generated by a 70 kOe split coil magnet, was rotated with a fine angular resolution of 0.01°. A representative set of data (ac-local magnetic permeability versus dc magnetic field for different angles) is shown in Fig. 1. In order to construct the phase boundary, we used the voltage level of 15μV (approximately middle point between zero and the saturation value of the real part of the permeability, μ).

3. Discussion

Comparison of the free energies of crossing and tilted vortex lattices result in a complicated picture of phase transitions between these vortex structures. The proposed phase diagram [4] is shown in Fig. 2.

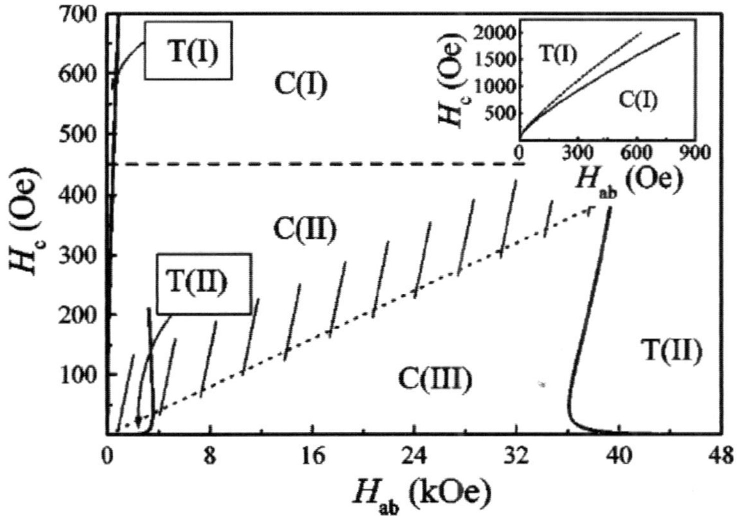

Figure. 2. The proposed phase diagram of the vortex-solid phase in the oblique magnetic fields [4].

The tilted vortex structure TI of inclined PV stacks can be replaced by the crossing lattice quite close to the *c* axis (see phase diagram obtained for γ = 500, inset in Fig. 2). The crossing lattice, which exists in a wide angular range, can have different substructures. At high enough out-of-plane fields, the ''shifted'' PV sublattice is realized in the crossing-lattices structure where the PVs are shifted from their equilibrium positions by currents generated by JVs.

In this substructure, the JV currents shift the PV's mostly along the *x* axis. The ''shifted'' phase can transform into the ''trapped'' PV lattice CII when the energy gain related to the ''crossing-lattice pinning'' exceeds the energy needed for the additional shear deformation (the dashed line in Fig. 2 separating CI and CII). Around the this line, the lattice CII can be changed by the tilted lattice TII with JV strings linked by PV kinks Fig. 2 bottom panel or by the crossing-lattice structure CIII at which all PV stacks are placed on a few JV's. As one can see, the vortex phase diagram in layered superconductors is very complicated even with no disorder or pinning. Disorder can affect this by both shifting phase boundaries from their no-disorder position and smear or even destroy phase boundaries. The experiments discussed here seems to be related to phase boundary between TII and CII (or CIII) phase boundary. This boundary is defined by the condition of the balance between the free energy F_t of the tilted lattice and the free energy of the crossing lattice F_c: $F_t(H_c,H_{ab}) = F_c(H_c,H_{ab})$. In the presence of pinning, we have to include the term, describing the interaction of vortices with disorder, which by itself is very non-trivial procedure. In the naive approach, we could just add a term F_p (H_c, H_{ab}, H_T), where H_T describes the dose of irradiation. One would expect the spreading $<[H_{ab}^{tran}(H_c)]^2> - <[H_{ab}^{tran}(H_c)]>^2$ of the phase line $H_{ab}^{tran}(H_c)$ to be proportional to H_T^2, which is consistent with our experimental results.

In conclusion, we study the deformation of vortex matter phase diagram of irradiated $Bi_2Sr_2CaCu_2O_{8+\delta}$ samples. The result is consistent with proposed earlier phase diagram for layered superconductor. However, a detailed description of disordered layered superconductors might require more detailed theoretical studies.

References

[1] Buzdin A I and Feinberg D 1990 *J. Phys. (Paris)* **51**, 1971
[2] Bulaevskii L and Clem J R 1991 *Phys. Rev. B* **44** 10234
[3] Koshelev A E 1999 *Phys. Rev. Lett.* **83** 187
[4] Savel'ev S E, Mirkovic J and Kadowaki K 2001 *Phys. Rev. B* **64** 94521
[5] Savel'ev S E, Mirkovic J and Kadowaki K 2001 *Physica C* **357–360** 597
[6] Savel'ev S E, Mirkovic J and Kadowaki K 2002 Physica C **378–381** 580
[7] Savel'ev S, Mirkovic J and Nori F 2003 *Physica C* **388** 653
[8] Tonomura A *et al* 2002 *Phys. Rev. Lett.* **88** 237001
[9] Matsuda T. *et al.*2001 *Science* **294** 2136
[10] Grigorenko A, Bending S, Tamegai T, Ooi S and Henini M.2001 *Nature* **414** 728
[11] Vlasko-Vlasov K, Koshelev A,Welp U, Crabtree G W and Kadowaki K 2002 *Phys. Rev. B* **66** 014523
[12] Cole D, Bending S, Savel'ev S, Grigorenko A, Tamegai T and Nori F 2006 *Nature Materials* **5** 305
[13] Savel'ev S and Nori F 2002 *Nature Materials* **1** 179
[14] Savel'ev S, Marchesoni F and Nori F 2004 *Phys. Rev. Lett.* **92** 160602
[15] Savel'ev S, Marchesoni F and Nori F 2003 *Phys. Rev. Lett.* **91** 010601
[16] Savel'ev S, Marchesoni F and Nori F 2004 *Phys. Rev. E* **70** 061107
[17] Savel'ev S, Marchesoni F and Nori F 2005 *Phys. Rev. E* **71**, 011107
[18] Savel'ev S, Marchesoni F, Hänggi P and Nori F 2004 *Eur. Phys. J. B* **40**, 403
[19] Mirkovic J, Savel'ev S E, Sugahara E and Kadowaki K 2001 *Phys. Rev. Lett.* **86**, 886
[20] Mirkovic J, Savel'ev S E, Sugahara E and Kadowaki K 2002 *Phys. Rev. B* **66** 132505
[21] Mirkovic J, Savel'ev S E, Sugahara E and Kadowaki K 2001 *Phys. Rev. Lett.* **86**, 886
[22] Mirkovic J, Savel'ev S E, Sugahara E and Kadowaki K 2000 *Physica C* **341** 1181
[23] Mirkovic J, Savel'ev S E, Sugahara E and Kadowaki K 2001 *Physica C* **357** 450
[24] Mirkovic J, Savel'ev S E, Sugahara, E. and Kadowaki K 2001 *Physica C* **364** 515
[25] Kadowaki K, Mirkovic J, Savel'ev S E, Sugahara E 2000 *Physica C* **341** 1301
[26] Tokunaga M, Kishi M, Kameda N, Itaka K and Tamegai T 2002 *Phys. Rev. B* **66** 220501 (R)
[27] Tachiki M, Iizuka M, Minami K, Tejima S and Nakamura H 2005 *Phys. Rev. B* **71** 134515
[28] Savel'ev S,Yampol'skii V, Rakhmanov A and Nori F 2005 *Phys. Rev. B* **72** 144515
[29] Savel'ev S,Yampol'skii V, Rakhmanov A. and Nori F. 2006 *Nature Phys.* **2** 521
[30] Savel'ev S, Rakhmanov A and Nori F. 2005 *Phys. Rev. Lett.* **94** 157004
[31] Savel'ev S,Yampol'skii V and Nori F 2006 *Phys. Rev. Lett.* **95** 187002
[32] Savel'ev S,Yampol'skii V and Nori F 2006 *Physica C* **445–448** 183
[33] Savel'ev S,Yampol'skii V, Rakhmanov A and Nori F 2010 *Rep. Prog. Phys.* **73**, 026501
[34] Hu X and Lin S Z 2010 *Supercond. Sci. Technol.* **23**, 053001
[35] Marchesoni F, Savel'ev S, Tachiki, M, Nori F 2010 *Phys. Rev. B* **81** 174531
[36] Ozyuzer L, Koshelev A E, Kurter C, Gopalsami N, Li Q, Tachiki M, Kadowaki K, Yamamoto T, Minami H,Yamaguchi H, Tachiki T, Gray K E, Kwok W K, Welp U 2007 *Science* **318**, 1291
[37] Minami H, Kakeya I, Yamaguchi H, Yamamoto T, Kadowaki K 2009 *Appl. Phys. Lett.* **95** 232511
[38] Tsujimoto M, Yamaki K, Deguchi K, Yamamoto T, Kashiwagi T,Minami H, Tachiki M, Kadowaki K, Klemm R A 2010 *Phys. Rev. Lett.* **105**, 037005

An energy competition of Co^{3+} and Co^{4+} ions during spin state transition in $Ca_3Co_4O_9$ complex

N Karaman[1], A Bayri[2], S Ekmekçi[3]

[1] Adiyaman University, Science and Art Faculty, Physics Department, 02040, Adiyaman, Turkey
[2] Inonu University, Science and Arts Faculty, Physics Department, 44280, Malatya, Turkey
[3] Adiyaman University, Education Faculty, Science Education Department, 02040, Adiyaman, Turkey

E-mail:nkaraman@adiyaman.edu.tr

Abstract. The magnetic behavior of $Ca_3Co_4O_9$ complex shows a very interesting phase transition at high temperature. It is probably that this ordering is due to the spin states of Co^{3+} and Co^{4+} ions. It is quite obvious that spin states of Co ions generally change as a function of temperature. We have shown that there is a strong correlation between oxidation states and ligand field parameters during spin state transitions.

1.Introduction

Magnetic molecular materials should have a key role to improve systems such as switches, optical displays, data storage devices and etc. Especially, switchable magnetic materials gained considerable interest in view of the potential technological applications after the spin crossover phenomenon (SCO) which was first discovered by Cambi et al [1]. Spin Crossover (SCO) is a phenomena on that occurs in some metal complexes where in the spin state of the complex changes due to external conditions such as, variation of temperature, pressure, light irradiation or even an influence of a field. Spin crossover behavior may be observed in compounds with transition metals of the first series with $3d^4$ to $3d^7$electron configurations and can be explained qualitatively by ligand-field theory [2].

In order to explain the effect of surroundings, there are two fundamental approaches namely Crystal Field Theory (CFT) and Ligand Field Theory (LFT). CFT is based upon a completely ionic picture of bonding. On the other hand LFT is based on covalent bonding. Figure 1 is a simple picture of how the degeneracy of d-orbitals are removed by the crystal field. When examining a single transition metal ion, the five d orbitals have the same energy. When we substituted it in a spherically symmetric field, the five d orbitals energies will be increased but the degeneracy will not change. When ligands approach the metal ion from different directions, all the energies of d electrons will be altered. For example, consider a molecule with octahedral geometry. Ligands approach the metal ion along the x ,y and z axes. Therefore, d_{z^2} and $d_{x^2-y^2}$ orbitals (which lie along these axes) are more effected compare with d_{xy}, d_{xz}, d_{yz} orbitals (which are oriented between x, y and z axes). This causes a splitting in the energy levels of the d-orbitals. This is known as crystal field splitting.

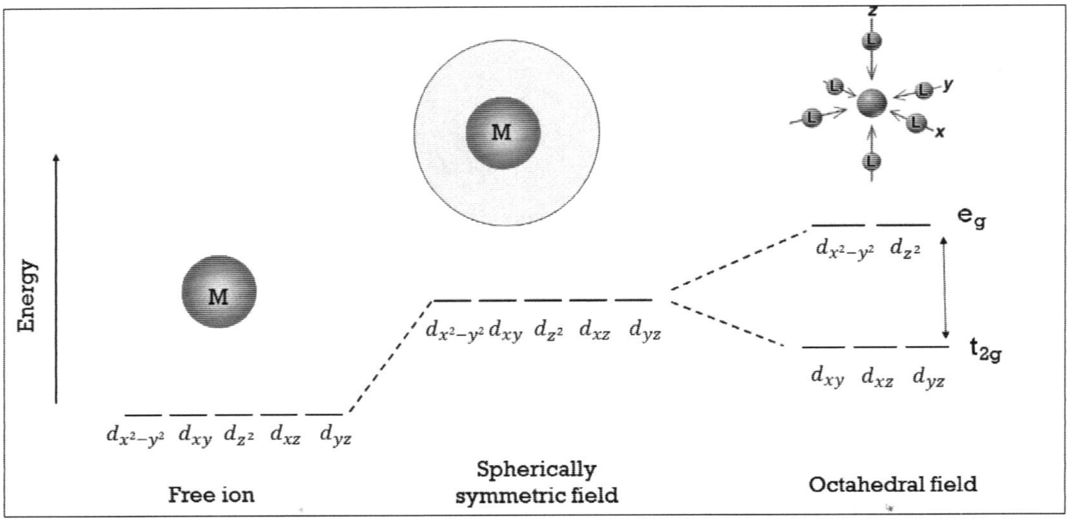

Figure 1. The five d orbitals lose their degeneracy by an octahedral arrangement of ligands.

For octahedral complexes, crystal field splitting energy (CFSE) is denoted by Δ_0. In an O_h point group,d_{z^2} and $d_{x^2-y^2}$ orbitals are called e_g orbitals and they belong to the E_g irreducible representation. Also in an O_h point group,d_{xy}, d_{xz}, d_{yz} orbitals are called t_{2g} orbitals and they belong to T_{2g} representation. The effects of ligand and crystal field are generally occurred in d^4-d^7 electrons. Up to three electrons, the configuration is same with atomic orbitals. On the other hand between d^4-d^7 electrons, configuration is generally different from atomic case. For example, as seen in Fig. 2, in d^3 configuration there would be 3 unpaired electrons. If we insert an extra electron in this configuration, there would be 2 possibilities because of strength of crystal field splitting energy. One of them is High Spin (HS) which has four unpaired electrons and the other is Low Spin (LS) configuration which has two unpaired electrons. The SCO phenomenon has uses in many applications due to the inherent bistability which leads to changes in the color of the material and major magnetic changes [3]. Molecular switches, like electrical switches, have the requirement for a mechanism that will turn the switch *on* and *off*. Figure 3 shows a schematic of the two different output signals that can be obtained from an Fe^{II} compound upon an external stimuli. In this study we have investigated magnetic properties of $Ca_3Co_4O_9$ complex which may be a candidate to be magnetic switch.

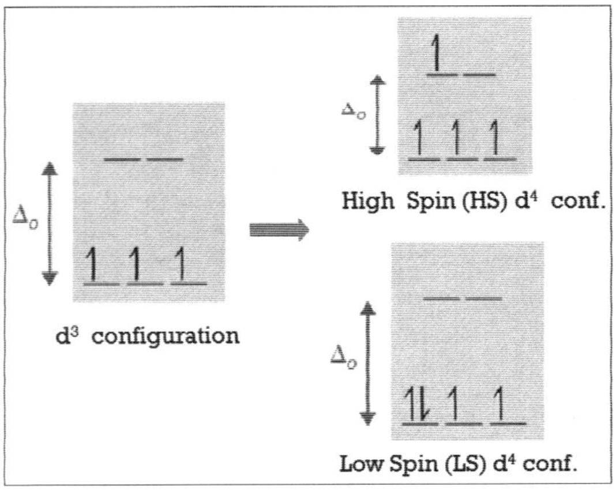

Figure 2. This is a figure with a caption that is wider than the actual graphic. To save space you can put the caption to the right of the figure by placing the graphic and justified caption in a table with one row and two columns.

2. Magnetic behavior of $Ca_3Co_4O_9$

Crystal structure of $Ca_3Co_4O_9$ consist of two subsystems and all Co ions in this structure have octahedral environment[4]. The magnetic moment of the sample obviously comes from Co ions. Co ions have three oxidation states in the complex; divalent (Co^{2+}), trivalent (Co^{3+}) and tetravalent (Co^{4+})[5]. All possible electronic configurations and symmetry term of Co ions are given in Figure 3.

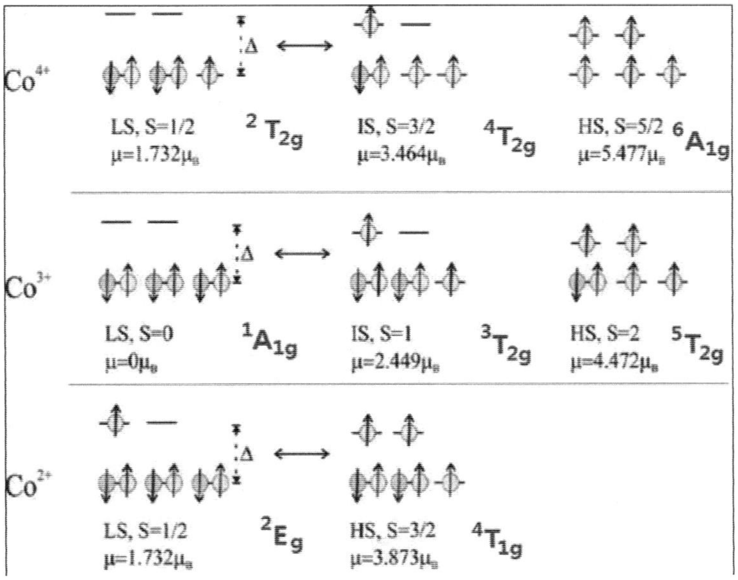

Figure 3. All possible electronic configurations and symmetry of Co ions.

Figure 4 shows temperature dependence of the magnetic susceptibility, in the temperature range of 2–1000 K. It is quite clear that there are two different spin state transition in the sample. First spin state transition is at about 380 K and second spin state transition intervals of 680 – 920 K and there is a sudden anomaly between 680 and 920 K [6]. We did two modeling in order to explain this anomaly and which Co ion will promote HS state earlier than other.

3. Models

3.1. Symmetry modeling

Co^{2+} ions have always the low spin state configuration. As seen in the figure 3, Co^{3+} ions make an ${}^1A_{1g} \rightarrow {}^5T_{2g}$ transition and Co^{4+}ionsmake a ${}^2T_{2g} \rightarrow {}^6A_{1g}$ transition when transitioned from LS to HS state. By the symmetry, *A* term is corresponding to spherically symmetric case. For this, *A* terms energy is smaller than the *T* term. Then, we have following suggestion; Co^{3+} ion needs,

$$\epsilon_1 = \Delta_{Co^{3+}} + E_1 \tag{1}$$

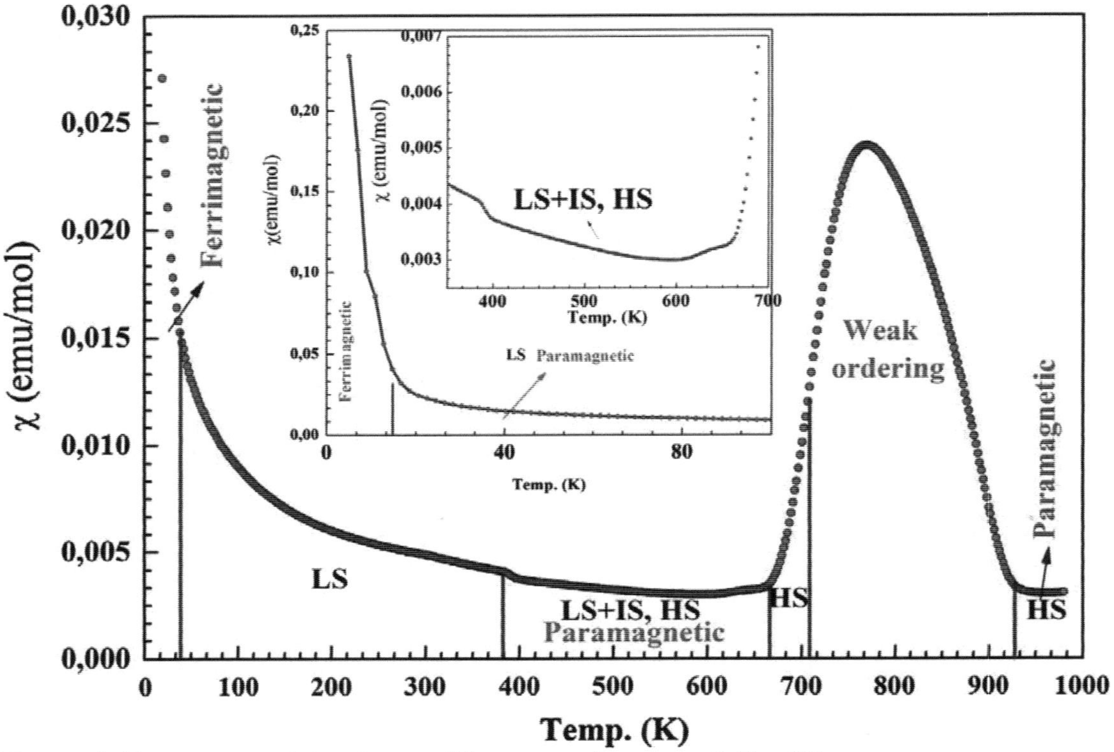

Figure 4. Temperature dependence of the magnetic susceptibility [6].

energy and Co^{4+}ion needs,

$$\epsilon_2 = \Delta_{Co^{4+}} - E_2 \quad (2)$$

Energy to make a transition from LS to HS state. In order to determine which ion would promote HS state, we must determine energy difference between (1) and (2) which is $\epsilon_1 - \epsilon_2$.

3.2 Oxidation state modeling

In order to determine the CFSE which depends on oxidation states of Co ions, we must determine orbitals energies. If we examining the Co^{2+} ion, it is a $3d^7$ electrons. First, if we consider $d_{x^2-y^2}$ orbital and ligands (as seen in figure 5). Then coulomb interaction energy between electrons in orbital and ligands will be $\delta e/r$, where r is distance between the electrons in orbital and ligands, δ is charge density of orbitals. Thus, $d_{x^2-y^2}$orbitals energy, $E_{d_{x^2-y^2}}(Co^{2+})$ is,

$$E_{d_{x^2-y^2}}(Co^{2+}) = atomic\ energy + \frac{\delta e}{r} \quad (3)$$

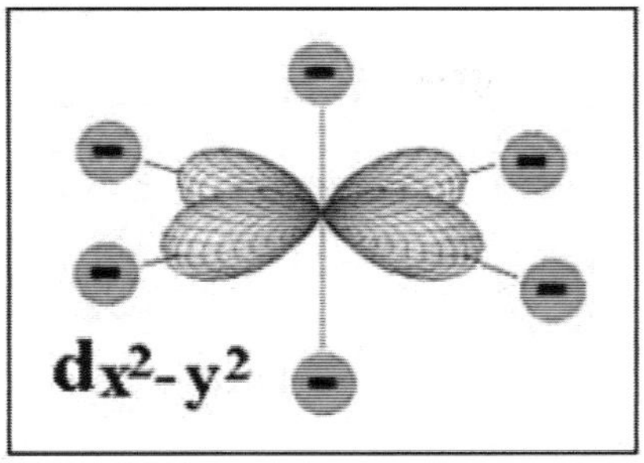

Figure 5.
$d_{x^2-y^2}$ orbital and ligands.

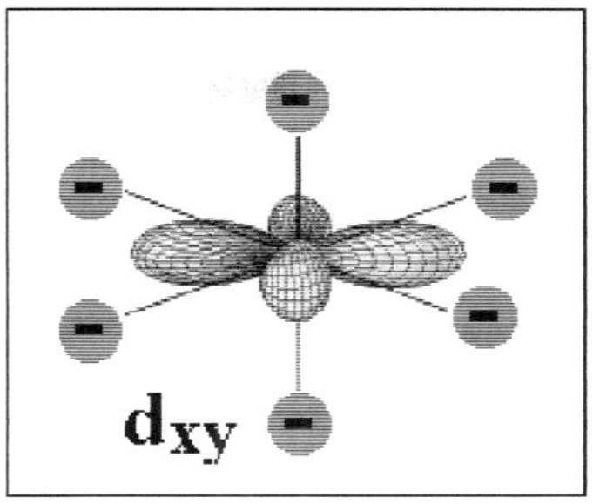

Figure 6.
d_{xy} orbital and ligands.

Second, if we consider d_{xy} orbital and ligands (as seen in figure 6); then coulomb interaction energy between electrons in orbital and ligands will be $\delta e/_{r'}$, where r' is distance between the electrons in orbital and ligands, δ is charge density of orbitals. Thus, d_{xy} orbitals energy $E_{d_{xy}}(Co^{2+})$ is,

$$E_{d_{xy}}(Co^{2+}) = atomic\ energy + \frac{\delta e}{r'} \tag{4}$$

CFSE of Co^{2+}ion is energy differences between orbitals energies:

$$\begin{aligned}\Delta_{Co^{2+}} &= E_{d_{x^2-y^2}}(Co^{2+}) - E_{d_{xy}}(Co^{2+}) \\ &= \delta e\left(\frac{1}{r} - \frac{1}{r'}\right)\end{aligned} \tag{5}$$

We put a similar argument to determine the CFSE of Co^{3+} ion where as we did Co^{2+} ion.

$$\begin{aligned}\Delta_{Co^{3+}} &= E_{d_{x^2-y^2}}(Co^{3+}) - E_{d_{xy}}(Co^{3+}) \\ &= \delta' e\left(\frac{1}{r_1} - \frac{1}{r_1'}\right)\end{aligned} \tag{6}$$

Where r_1 is distance between the electrons in $d_{x^2-y^2}$ orbital and ligands, r_1' is distance between the electrons in d_{xy} orbital and ligands and δ' is charge density of orbitals for Co^{3+}ion. When a Co^{2+} ion loses an electron, it oxidizes to Co^{3+} ion and effective nuclear charge is increased. So ligands are closer to metal ion and repulsion between the d orbitals and ligands will be greater [7]. Namely; $r_1 < r$ and $\delta' > \delta$ then $E_{d_{x^2-y^2}}(Co^{3+}) > E_{d_{x^2-y^2}}(Co^{2+})$ and also $r_1' < r'$ and $\delta' > \delta$ then $E_{d_{xy}}(Co^{3+}) > E_{d_{xy}}(Co^{2+})$. For these reasons $\Delta_{Co^{3+}}$ is greater than $\Delta_{Co^{2+}}$. For similar reasons, we can say $\Delta_{Co^{4+}} > \Delta_{Co^{3+}}$. In order to determine which ion would promote to HS state, we must determine CFSE difference between $Co^{3+}$and Co^{4+}ions; $\Delta_{Co^{4+}} - \Delta_{Co^{3+}}$.

4. Conclusion

We considered two different interaction models, namely *the symmetry effect* and *oxidation state*. It is obvious that symmetry prefers Co^{4+} ions promote to HS state earlier than Co^{3+} ions. On the other hand the oxidation state prefers earlier promotion of Co^{3+} ions. It is certain that there is a competition between the set effects in $Ca_3Co_4O_9$ sample.

References

[1] Cambi L and Szegö L1931 *Eur. J. Inorg. Chem.* **64** (10)2591–2768
[2] Boca R 1999 *Theoretical Foundations of Molecular Magnetism* (Amsterdam: Elsevier) chapter 9 pp 541-578
[3] Gütlich P and Goodwin H A 2004 *Spin Crossover in Transition Metal Compounds I* (Berlin: Springer)
[4] Paulauskas T and Klie R F 2011 *J. of Undergraduate Research* **5** 1
[5] Masset A C and Michel H. 2000 *Physical Review B* **62** 166-174
[6] Altin S and Aksan M A 2014 *J.of Alloys and Compounds* **587** 40-44
[7] Figgs B N and Hitchman M A 2000 *Ligand Field Theory and its Applications* (New York: Wiley-VCH)chapter 7 pp 163-166

Generation of uniform magnetic field using a spheroidal helical coil structure

Yavuz Öztürk[1,2] and Bekir Aktaş[1]

[1]Gebze Technical University (GTU), Physics Department, Gebze/Kocaeli, Türkiye.
[2]TÜBITAK BILGEM, Gebze/Kocaeli, Türkiye.

E-mail: yavuz.ozturk@tubitak.gov.tr

Abstract. Uniformity of magnetic fields are of great importance especially in magnetic resonance studies, namely in magnetic resonance spectroscopy applications (NMR, FMR, ESR, EPR etc.) and magnetic resonance imaging applications (MRI, FMRI). Field uniformity is also required in some other applications such as eddy current probes, magnetometers, magnetic traps, particle counters etc. Here we proposed a coil winding regime, which follows the surface of a spheroid (an ellipsoid of rotation); in light of previous theoretical studies suggesting perfect uniformity for a constant ampere per turn in the axial direction thereof. We demonstrated our theoretical results from finite element calculations suggesting 0.15% of field uniformity for the proposed structure, which we called a Spheroidal Helical Coil.

1. Introduction

Maintaining high magnetic field uniformity is the key feature for many magnetic monitoring applications. Magnetic Resonance Imaging (MRI) instruments need extreme field intensities (over 1 Tesla) with high field uniformity. The former is required for the sake of increasing signal-to-noise ratio of the weak resonance signal without increasing the acquisition rate beyond reasonable time scales. The latter is needed to address the signals from the body under examination. Since using permanent magnets is not practical both in the sense of manufacturing and utilization of the device, magnetic fields are generated using coils in typical imaging machines. Modern MRI instruments depend on superconducting wires and cryogenic technology to reach extreme intensities, and shimming techniques to maintain high field uniformity [1].

Helmholtz coils are the most practical and widely used winding structures for obtaining field uniformity in a relatively small region [2,3]. Alternative coil structures like, tetra coils [4,5], saddle coils [6,7] or planar coils [8] are also used to reach high field uniformity. Ellipsoidal structures, which offer theoretically perfect uniform magnetic fields, are studied as well. J. C. Maxwell was maybe the first to work out the demagnetization constants of ellipsoidal structures [9]. Later, E. C. Stoner [10] and J. A. Osborn [11] investigated demagnetization in ellipsoidal structures and published detailed analysis and tables in their respective articles. Similar calculations can also be seen in the famous books by J. A. Stratton [12] and L. D. Landau [13]. In his paper "on the problem of the ellipsoid" G. E. Marsh showed that a linear current function defined on the ellipsoidal surface can generate a homogeneous magnetic field inside along its principle axis [14].

2. Theory

In case the so-called current function (Φ) is linear along the z axis, it will have the form:

$$\Phi = -Kz \tag{1}$$

Marsh proved that this assumption in (1) will result in a homogeneous field inside an ellipsoid [14]. The relation between the current density and the current function is proven to be:

$$\boldsymbol{j} = \nabla\Phi \times \hat{\boldsymbol{n}} \tag{2}$$

where $\hat{\boldsymbol{n}}$ is normal to the surface of the ellipsoid [14].

The derivation of the magnetic field is started using the discontinuity of the vector potential ($\boldsymbol{A}$) at the boundary (the surface):

$$\frac{\partial \boldsymbol{A}^+}{\partial n} - \frac{\partial \boldsymbol{A}^-}{\partial n} = -\mu_0 \mathbf{j} \tag{3}$$

The vector potential also obeys the Laplace equation ($\nabla^2 A_i = 0$).The Laplace equation in ellipsoidal coordinates for $\boldsymbol{A}$ is as follows:

$$(\eta - \zeta)R_\lambda \frac{\partial}{\partial\lambda}\left(R_\lambda \frac{\partial\phi}{\partial\lambda}\right) + (\zeta - \lambda)R_\eta \frac{\partial}{\partial\eta}\left(R_\eta \frac{\partial\phi}{\partial\eta}\right) + (\lambda - \eta)R_\zeta \frac{\partial}{\partial\zeta}\left(R_\zeta \frac{\partial\phi}{\partial\zeta}\right) = 0 \tag{4}$$

where

$$R_\lambda = [(a^2+\lambda)(b^2+\lambda)(c^2+\lambda)]^{1/2} \tag{5}$$

In case the family of confocal ellipsoids –which are parameterized by λ - are equipotential lines, one must have:

$$\frac{\partial}{\partial\lambda}\left(R_\lambda \frac{\partial\phi}{\partial\lambda}\right) = 0 \tag{6}$$

From here Marsh showed that, $\phi_0 = \text{const} \cdot \int_0^\infty d\lambda / R_\lambda$ being the elliptic integral and $\hat{\boldsymbol{z}}$ being the unit vector along the z direction, the magnetization can be derived to be:

$$\boldsymbol{B} = \mu_0 abcK\left(\frac{d\phi_0}{da^2} + \frac{d\phi_0}{db^2}\right)\hat{\boldsymbol{z}} \tag{7}$$

which is constant along z depending only on ellipsoidal parameters [14].
When the current is calculated by using the $\Phi = -Kz$ current function, one can write down:

$$\boldsymbol{j} = \nabla\Phi \times \hat{\boldsymbol{n}} = -K(\nabla z) \times \hat{\boldsymbol{n}} = K(\hat{\boldsymbol{n}} \times \hat{\boldsymbol{z}}) \tag{8}$$

From here we may conclude that in order to have constant current, $(\hat{\boldsymbol{n}} \times \hat{\boldsymbol{z}})$ has to be constant as well. Since $|\hat{\boldsymbol{n}} \times \hat{\boldsymbol{z}}|$ is also constant, $\hat{\boldsymbol{n}} \cdot \hat{\boldsymbol{z}} = constant \equiv d$ can be defined. Thus one can obtain

$$\frac{x^2}{a^4}+\frac{y^2}{b^4}-\left(\frac{1}{d^2}-1\right)\frac{z^2}{c^4}=0 \tag{9}$$

which is the equation of a real quadric cone. The intersection of this ellipsoid with the unit ellipsoid

$$\frac{x^2}{a^2}+\frac{y^2}{b^2}+\frac{z^2}{c^2}=1 \tag{10}$$

is the desired winding curve. So using (10) to eliminate z in (9) one will have:

$$\left(\frac{c^2}{a^2}\frac{d^2}{1-d^2}+1\right)\frac{x^2}{a^2}+\left(\frac{c^2}{b^2}\frac{d^2}{1-d^2}+1\right)\frac{y^2}{b^4}=1 \tag{11}$$

This is actually the equation of an ellipse with semi-axes:

$$a\left(\frac{c^2}{a^2}\frac{d^2}{1-d^2}+1\right)^{-1/2} \quad \text{and} \quad b\left(\frac{c^2}{b^2}\frac{d^2}{1-d^2}+1\right)^{-1/2} \tag{12}$$

From here, the introduction of $\hat{\boldsymbol{n}}\cdot\hat{\boldsymbol{z}}=\cos\theta$ and parameterization of the ellipsoid with

$$\begin{aligned} x^2 &= a^2\sin^2\alpha\cos^2\beta, \\ y^2 &= a^2\sin^2\alpha\sin^2\beta \\ z^2 &= a^2\cos^2\alpha \end{aligned} \tag{13}$$

yield

$$\cot^2\alpha = c^2\cot^2\theta\left(\frac{1}{a^2}+\left(\frac{1}{b^2}-\frac{1}{a^2}\right)\sin^2\beta\right) \tag{14}$$

Here for the case of a spheroid ($a = b$),$z$variation must vanish, thus the parameter αhas to be a constant giving

$$\cot^2\alpha=\frac{c^2}{a^2}\cot^2\theta=\frac{c^2}{a^2}\frac{1-\sin^2\theta}{\sin^2\theta}=\frac{c^2}{a^2}\left(\frac{1}{\sin^2\theta}-1\right) \tag{15}$$

This corresponds to the case of a spheroid where the internal field will be uniform if the current is proportional to:

$$I \propto \sin\theta=\left(1+\frac{a^2}{c^2}\cot\alpha\right)^{-1/2} \tag{16}$$

This is equivalent, in terms of windings, to having a constant number of ampere per turns per unit length along the z axis [14]. This result is also confirmed by the earlier studies of L. J. Laslett [15]and J. E. Everett [16] for the special case of a sphere. A similar study was pursued by Živaljević and Aleksić on prolate and oblate spheroids. They approached the problem by defining a scalar magnetic potential for solving the Laplace equation using the suitable boundary conditions [17]. They have used both prolate and oblate spheroidal coordinates respectively (Figure 1).

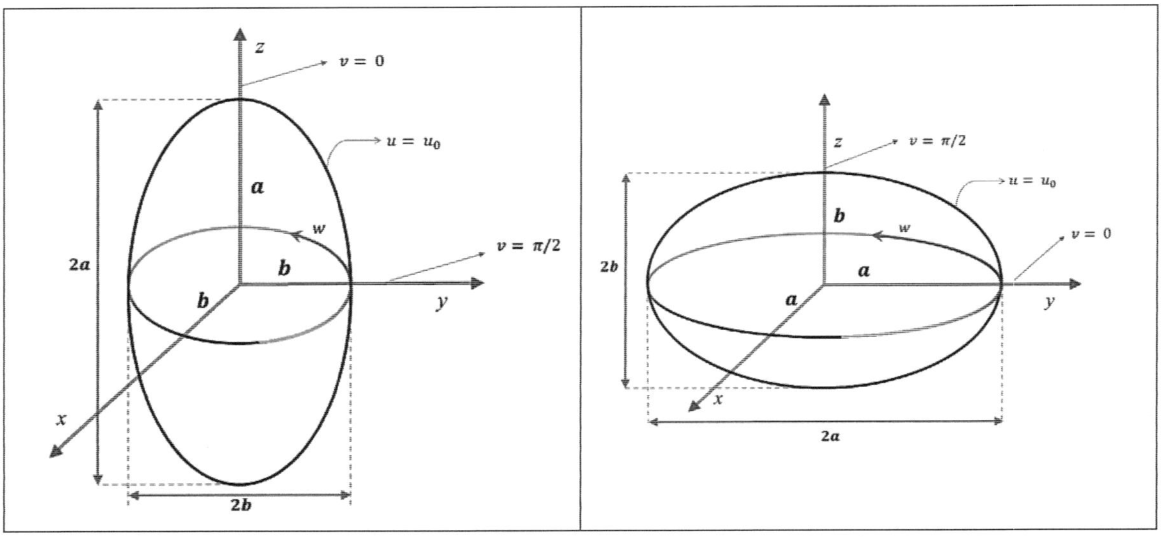

Figure 1. Prolate (left) and oblate (right) ellipsoidal coordinate definitions.

Using prolate spheroidal coordinates they wrote down the electric and magnetic field vectors as follows:

$$\boldsymbol{E}(u,v) = E_w(u,v)\widehat{\boldsymbol{w}} \tag{17}$$

$$\boldsymbol{H}(u,v) = H_u(u,v)\widehat{\boldsymbol{u}} + H_v(u,v)\widehat{\boldsymbol{v}} \tag{18}$$

where $\widehat{\boldsymbol{w}}, \widehat{\boldsymbol{u}}, \widehat{\boldsymbol{v}}$ are prolate spheroidal coordinate vectors and in terms of Cartesian unit vectors:

$$\begin{aligned} \widehat{\boldsymbol{u}} &= \frac{c}{h_u}(\cosh u \,\sin v \,\cos w \,\widehat{\boldsymbol{x}} + \cosh u \,\sin v \,\sin w \,\widehat{\boldsymbol{y}} + \sinh u \,\cos v \,\widehat{\boldsymbol{z}}) \\ \widehat{\boldsymbol{v}} &= \frac{c}{h_v}(\sinh u \,\cos v \,\cos w \,\widehat{\boldsymbol{x}} + \sinh u \,\cos v \,\sin w \,\widehat{\boldsymbol{y}} + \cosh u \,\sin v \,\widehat{\boldsymbol{z}}) \\ \widehat{\boldsymbol{w}} &= -\sin w \,\widehat{\boldsymbol{x}} + \cos w \,\widehat{\boldsymbol{y}} \end{aligned} \tag{19}$$

Here *h* factors are called Lame coefficients and for prolate coordinates they are defined as:

$$\begin{aligned} h_u = h_v = h &= c(\sinh^2 u + \sin^2 v)^{1/2} \\ h_w &= c \sinh u \sin v \end{aligned} \tag{20}$$

where a and b are elliptic constants for $a > b$ and $c = \sqrt{a^2 - b^2}$ is the eccentricity [18]. Solving the Laplace equation, the magnetic field is found to be:

$$H(u,v) = C_0 J_s(\cot v \sinh u \,\hat{u} - \cosh u \,\hat{v}) \tag{21}$$

where C_0 is a constant of spheroid and equals to:

$$C_0 = \frac{\sinh^2 u_o}{2} \ln \frac{\cosh u_0 + 1}{\cosh u_0 - 1} - \cosh u_0 \tag{22}$$

while J_s is the surface current [17]. Using (19) one may write down:

$$\boldsymbol{H}(u,v) = \frac{cC_0J_s}{h}(\cosh u \sinh u \cos v \cos w - \cosh u \sinh u \cos v \cos w)\hat{\boldsymbol{x}} + \frac{cC_0J_s}{h}(\cosh u \sinh u \cos v \cos w - \cosh u \sinh u \cos v \cos w)\hat{\boldsymbol{y}} + \frac{cC_0J_s}{h \sin v}(\cos^2 v \sinh^2 u - \sin^2 v \cosh^2 u)\hat{\boldsymbol{z}} \tag{23}$$

$$\boldsymbol{H}(u,v) = \frac{cC_0J_s}{h \sin v}\big((1 - \sin^2 v)\sinh^2 u - \sin^2 v \cosh^2 u\big)\hat{\boldsymbol{z}}$$

$$\boldsymbol{H}(u,v) = \frac{cC_0J_s}{h \sin v}(\sinh^2 u + \sin^2 v\,(\cosh^2 u - \sinh^2 u))\hat{\boldsymbol{z}} \tag{24}$$

$$\boldsymbol{H}(u,v) = \frac{cC_0J_s}{h \sin v}(\sinh^2 u + \sin^2 v)\hat{\boldsymbol{z}}$$

$$\boldsymbol{H}(u,v) = \frac{C_0J_sh}{c \sin v}\hat{\boldsymbol{z}}$$

From here it is deduced that the surface current (J_s) has to be of the form:

$$J_s \propto \sin v/h \tag{25}$$

in order to have uniform magnetic field inside the spheroid. Consequently the magnetic field inside will be:

$$\boldsymbol{H} = \frac{C_0J_0}{\sqrt{a^2 - b^2}}\hat{\boldsymbol{z}} \tag{26}$$

where J_0 is the constant of proportionality in (25) [17]. Živaljević and Aleksić also calculated the magnetic field outside the spheroid.

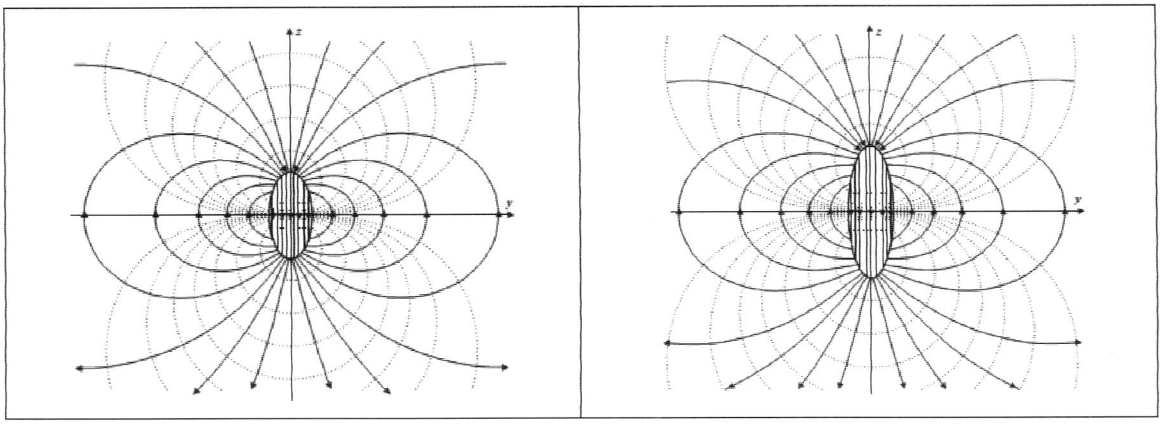

Figure 2. Prolate ellipsoidal coil magnetostatic field distribution and equipotential lines for $\boldsymbol{a/b = 2}$ and $\boldsymbol{a/b = 3}$ ratios respectively [17].

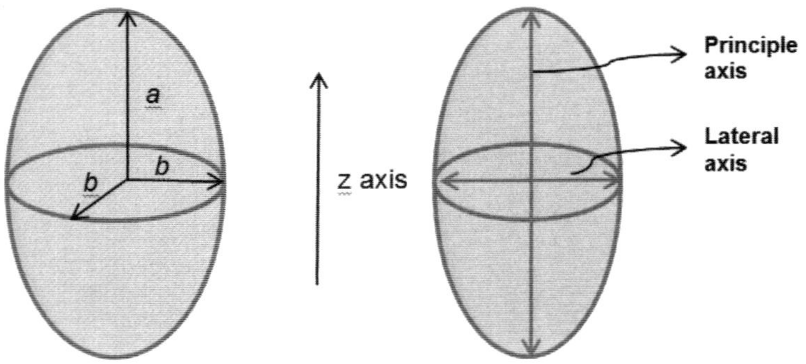

Figure 3. **The spheroid dimensions and axes.**

Their results for both inside and outside for $a/b = 2$ and $a/b = 3$ can be seen in Figure 2. As a further step using (25) they deduced a winding function N (a coil number density) to be as follows:

$$N = \frac{N_0 \sin v}{h} \quad (27)$$

Their results (as in Figure 2) are consistent with the previous literature [9-12], however the so-called winding function does not say much about the actual winding of the coil. In fact it can be shown that the result in equation (25) can produce the current function of Marsh [14] in equation (1). Hence the derivations made by Živaljević and Aleksić [17] can be reconciled with the well-known constant ampere per turn result. The derivation connecting the number density in (27) to the current function in (1) is carried out elsewhere [19].

3. Calculations and results

In order to confirm the conclusion of "constant number of ampere per turns per unit length along the z axis", we set up some Finite Element Analysis (FEA) calculations using CST EM Studio© software. We defined a hundred current rings of which radii obeying a spheroid having dimensions$a = 320\ mm$ and $b = 160\ mm$. A unit ampere of current was assumed to flow through each current carrying ring. The rings were defined to be of copper wires 0.9 mm in diameter. Simulation results were plotted on the principle (z) and lateral axes (which can be taken as x axis) (Figure 3). Defined structure and cross-section of a few meshed wires can be seen in Figure 4. Our results from FEA calculations along principle and lateral axes are given in Figure 5 and Figure 6, respectively. Field uniformities along both axes were calculated to be 0.03%. Clearly seen that these results confirm the theory.

That kind of a ring assembly is not fully realizable, unless one has current preserving superconducting rings. Yet there have been some experimental works applying similar methods. Since current has to be supplied to the system, there has to be some input and output junctions. Namely, current has to be introduced to the coils and has to leave them by some means. In order to avoid using multiple current supplies, open ring structures which are progressively soldered to each other were preferred in previous works [16,19]. On the other hand, these works were addressing respectively large structures where the effects and defects from solders and jumpers can be omitted.

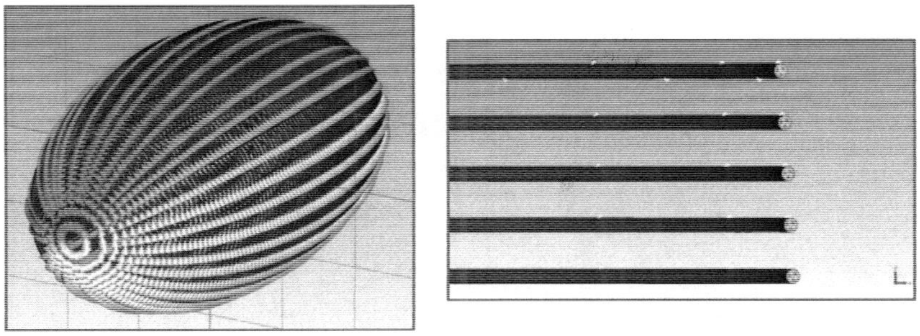

Figure 4. Modeled spheroidal current rings structure is on the left. Small arrows symbolize the current flow on the rings. Cross-sectional view of the meshed circular rings is on the right. The rings are defined to be of copper wires having 0.9 mm of diameter.

For small scale structures this method is expected to sensibly affect the magnetic field uniformity. In order to attain an experimentally feasible structure in which the field uniformity can be maintained, we proposed a coil winding regime which was described by the following formulas.

$$x = b\sqrt{1-\frac{z^2}{a^2}}\cos\left(\frac{2\pi z}{d}\right)$$

$$y = b\sqrt{1-z^2/a^2}\,\sin(2\pi z/d) \tag{28}$$

where $a > b, -a \leq z \leq a$ and d is the z distance between windings and calculated by dividing the principle axis length to the desired number of windings ($d = 2a/N_{windings}$). The coil structure proposed above is a *spheroidal helical coil* structure.

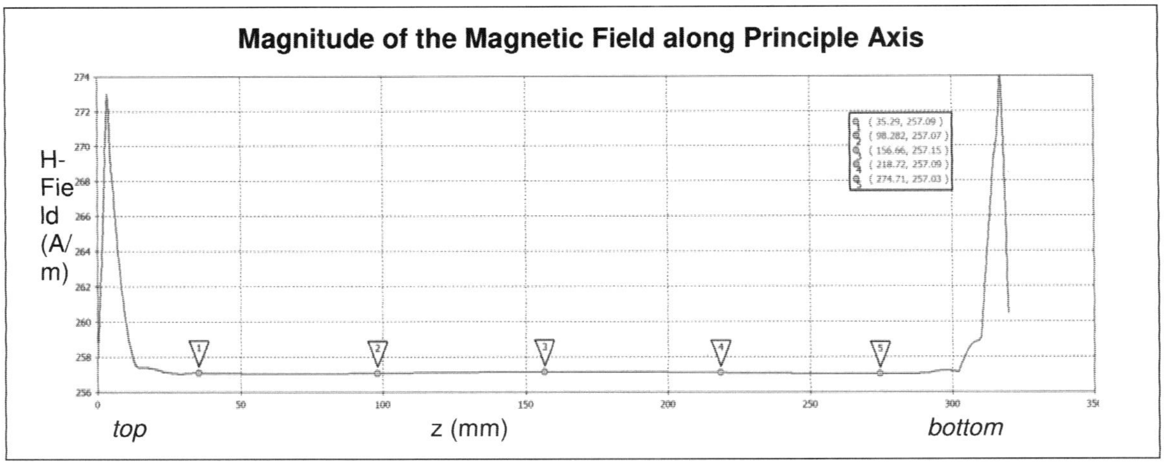

Figure 5. Magnitude of the H-field along principle axis is plotted. The principle field uniformity was calculated to be 0.03%.

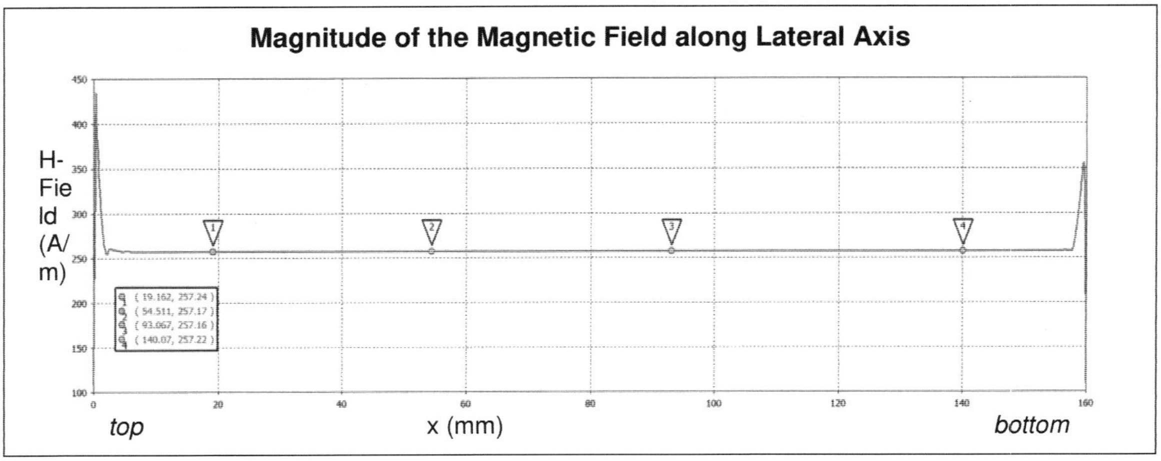

Figure 6. Magnitude of the H-field along lateral axis is plotted. The lateral field uniformity was calculated to be 0.03%.

Calculations were conducted using CST EM Studio© as before and the coil was again accepted to be of a 0.9 mm copper wire. Current to the coil was introduced by a current path which was assumed to be a perfect wire. The path curled around the coil far enough not to influence the results. Our results from FEA calculations along principle and lateral axes can be seen in Figure 8 and 9 respectively. Field uniformities along both axes were calculated to be %0.15 which are considerably successful results. The spikes at the ends of both figures were considered to be artifacts caused partly by the inconvenience of meshing procedure, which was not good enough to resolve the magnetic field just above the coils. Another reason may be the slight mismatch between the coil and the current path which introduced artificial discontinuities in the first derivative of the current.

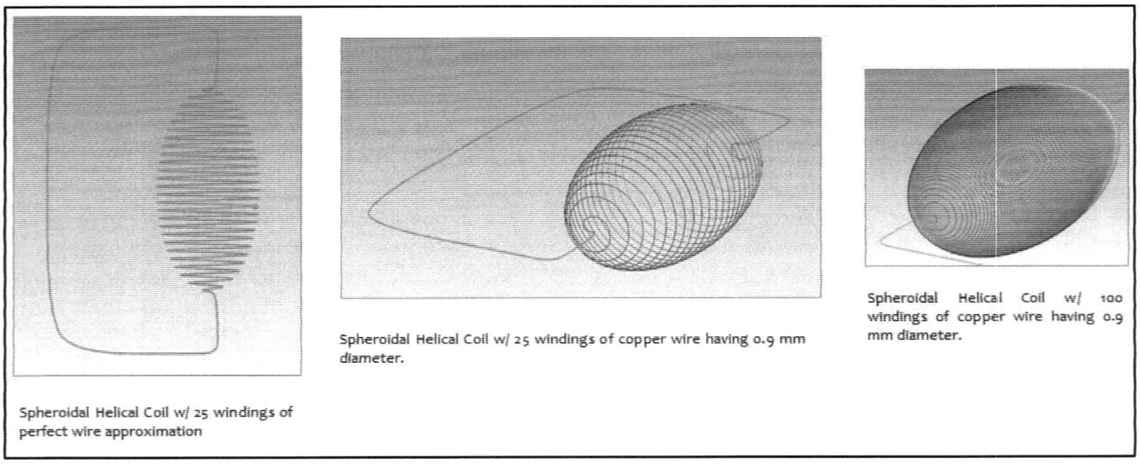

Spheroidal Helical Coil w/ 25 windings of perfect wire approximation

Spheroidal Helical Coil w/ 25 windings of copper wire having 0.9 mm diameter.

Spheroidal Helical Coil w/ 100 windings of copper wire having 0.9 mm diameter.

Figure 7. Some spheroidal helical coil structures.

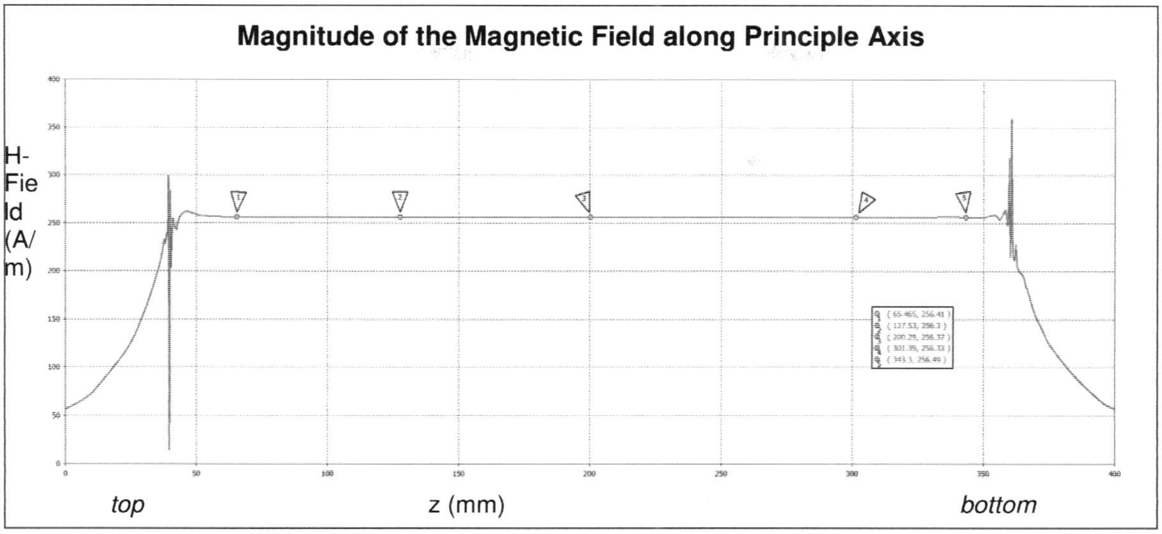

Figure 8. Magnitude of the H-field along principle axis for the spheroidal helical coil is plotted. The principle field uniformity was calculated to be 0.15%.

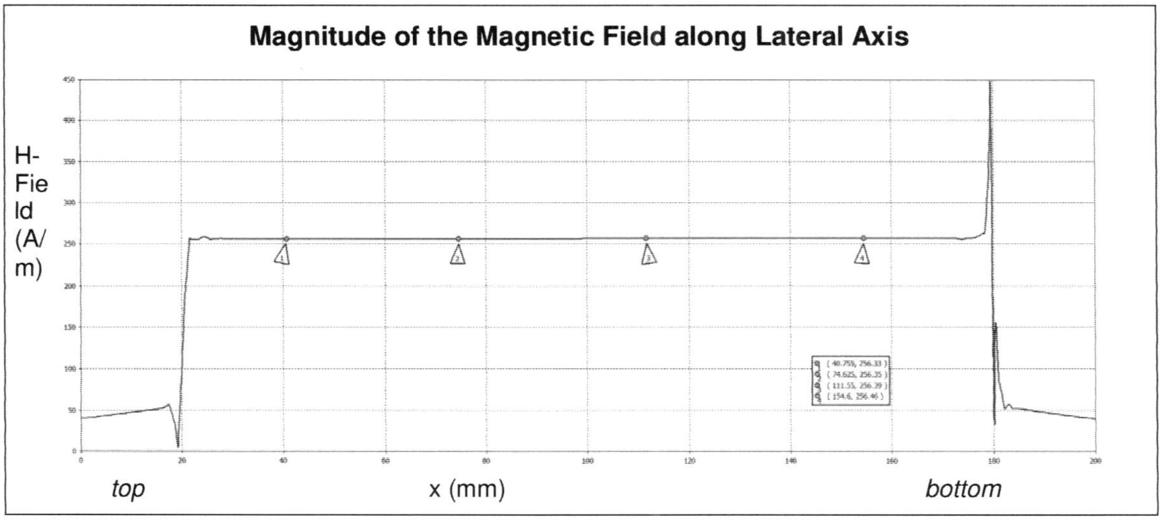

Figure 9. Magnitude of the H-field along lateral axis for the spheroidal helical coil is plotted. The lateral field uniformity was calculated to be 0.15%.

4. Conclusion

In this work, theoretical studies for ellipsoidal structures are shortly reviewed and their results which showed constant number of ampere per turns per unit length along the z axis are confirmed via FEA calculations. As a further step a continuous winding structure on an ellipsoid of rotation – which we called spheroidal helical coil structure – have been proposed and also confirmed by FEA calculations and proven to be a good candidate for the generation of a uniform magnetic field within a closed volume.

References

[1] Kuperman V 2000 *Magnetic Resonance Imaging: Physical Principles and Applications* (San Diego: Academic Press) p 9–26
[2] Magdaleno-Adame S, Olivares-Galvan J C,Campero-Littlewood E, Escarela-Perez R, Blanco-Brisset E 2010 *Coil Systems to Generate Uniform Magnetic Field Volumes* Excerpt from the Proceedings of the COMSOL Conference 2010 Boston
[3] Bell G B and Marino A A 1989 Exposure systems for production of uniform magnetic fields *J. of Bioelectricity* **8 (2)** 147–58
[4] Merritt R, Purcell C and Stroink G 1983 Uniform magnetic field produced by three, four, and five square coils *Rev. Sci. Instrum.* **54** 879–82
[5] Gottardi G, Mesirca P, Agostini C, Remondini D and Bersani F, 2003 A four coil exposure system (tetracoil) producing a highly uniform magnetic field *Bioelectromagnetics* **24** 125–33
[6] Bonetto F, Anoardo E and Polello M 2006 Saddle coils for uniform static magnetic field generation in NMR experiments *Concepts Magn. Reson.* Part B **29 B1** 9–19
[7] Dinale J and Vrbancich J 2014 Generation of long prolate volumes of uniform magnetic field in cylindrical saddle-shaped coils *Meas. Sci. Technol.***25** 035903 1–13
[8] Sasada I and Nakashima Y 2006 Planar coil system consisting of three coil pairs for producing a uniform magnetic field *J. Appl. Phys.* **99** 08D904
[9] Maxwell J C 1904 *Electricity and Magnetism* ed 3 vol 2 (Oxford: The Clarendon Press) p 66–70
[10] Stoner E C1945 Demagnetizing factors for ellipsoids *Phil. Mag.* **36** p 803–21
[11] Osborn J A 1945 Demagnetizing factors of the general ellipsoid *Phys. Rev.* **6** 7352–57.
[12] Stratton J 1941 *Electromagnetic Theory* (McGraw-Hill Book Company, Inc.: New York) p 209–13
[13] Landau L D and Lifshitz E M 1960 *Electrodynamics of Continuous Media* (Oxford: Pergamon Press) p 20–27
[14] Marsh G E1975 Uniform magnetic field in deflection coil design and the problem of ellipsoid *J. Appl. Phys.* **46** 3178–81
[15] Laslett L J 1966 An equivalent distribution of surface currents for the generation of a prescribed static magnetic field within the enclosed volume *J. Appl. Phys.* **37** 2361–63
[16] Everett J E and Osemeikhian J E 1966 Spherical coils for uniform magnetic fields *J. Sci. Instrum.* **43** 470–74
[17] Živaljevič D U and Aleksič S R 2007 Generating homogeneous magnetostatic field inside prolate and oblate ellipsoidal coil *Int. J. Electron. Commun. (AEÜ)* **61** 637–44
[18] Arfken G 1970 *Mathematical methods for physicists (2nd ed.)* (New York: Academic Press) p 103–8
[19] Öztürk Y and Aktaş B *to be published*
[20] Clark J W 1938 A new method for obtaining a uniform magnetic field. *Rev. Sci. Instrum.* **9** 320–22

Journal of Physics: Conference Series **667** (2016) 012010 doi:10.1088/1742-6596/667/1/012010

The effects of thermal procedure on transformation temperature, crystal structure and microstructure of Cu-Al-Co shape memory alloy

Y Aydoğdu[1], M Kök[2], F Dağdelen[2], A Aydoğdu[1], A S Turabi[3] and H E Karaca[3]

[1]Gazi University, Faculty of Science, Department of Physics, Superconductivity and Thermal Analysis Laboratory, Ankara, Turkey
[2]Firat University, Faculty of Science, Department of Physics, Elazig, Turkey
[3]Department of Mechanical Engineering, University of Kentucky, Lexington, Kentucky, 40506, USA

E-Mail:y.aydogdu@gazi.edu.tr

Abstract. The purpose of this study is to investigate the effects of different thermal procedures of the Cu-Al-Co shape memory alloy on its crystal structure, transformation temperature and microstructure. The alloys were subjected to a heat treatment and then cooling was applied at four different conditions. After the thermal process, XRD, DSC, optical microscopy and micro-hardness measurements were carried out. The experimental studies showed that crystal structure, microstructure and transformation temperature of Cu-Al-Co alloy were changed from the cooling conditions.

1.Introduction

There is a significant demand on Cu-based shape memory alloys for high temperature applications as these alloys have martensite transformation temperatures higher than 200 °C [1]. Addition of a third element into the Cu-Al binary alloys represents the regular DO_3 structure in the form of $(CuX)_3Al$. Here, Cu atoms are substituted by X atoms, where Cu atoms are dominant. If the amount of X is increased significantly, then the crystal structure transforms from DO_3 to $L2_1$ of which stoichiometric representation is Cu_2XAl [2]. Cu-Al-Co, which has been started to investigate recently, represents a group of high temperature shape memory alloys about which the knowledge is limited [3,4]. In the literature, generally, changes in the martensitic transformation of Cu-Al-Co alloy after the production and post-production stages have been investigated crystallographically and thermally [1,3]. However, there is lack of study focusing on the thermal procedures applied to these alloys.

Martensitic transformations of the most of the Cu-based alloys are affected from quenching and post quenching. More clearly, it can be stated that stabilization temperatures of these alloys depend on the quenching condition. Direct quenching from high temperature is quite effective to achieve full martensitic transformation [5,6]. Quenching is a different heat treatment process applied to the shape memory alloys. Generally, the most of the heat treatment procedures performed on the shape memory alloys are known as the application of a high temperature heat treatment after which the alloy is suddenly cooled by immersing it into a water with ice-salt bath. There are studies mainly concerning about the temperature of the heat treatment, while those dealing with the temperate of the cooling bath have been limited [5,6;7-11]. For this reason, in this study, the heat treatment temperature was kept constant at 850 °C, while the effects of different cooling media on martensitic transformation temperature, crystal structure, microstructure and micro-hardness of the Cu-Al-Co alloy were investigated.

2. Materials and methods

Cu-%at.21Al-%at.1Co shape memory alloy samples were cut from ingot such that they form five different groups. The processes outlined as follow:

a) M0; Without heat treatment
b) M1; Heat treatment at 850 °C for 1 hour and then quenching in water with ice-salt mixture
c) M2; Heat treatment at 850 °C for 1 hour and then quenching in water at room temperature
d) M3; Heat treatment at 850 °C for 1 hour and then quenching in boiling water for 3 min and finally quenching in water with ice-salt mixture
e) M4; Heat treatment at 850 °C for 1 hour and then furnace cooled.
The cooling rate of the alloys could be ranked as M1>M2>M3>M4. These are the results observed in the course of the experimental processes.

DSC measurements of the Cu-Al-Co alloys were conducted between 100 and 450 °C at a cooling rate of 25 °C/min under the nitrogen atmosphere so as to investigate the effects of the cooling rate and cooling conditions on the transformation temperature. Similarly, XRD measurements were performed by Bruker Discover D8 computer X-Ray diffractometer to identify the changes occurred in the crystal structure. Finally, optical microscopy investigation of the samples, which were moulded into polyester, were performed by using etchant made out of 20 ml HCl, 5 g $FeCl_3$-H_2O and 96 ml methanol and Nikon Eclipse MA200 optical microscope and the micro-hardness measurements of the same samples were conducted by a micro-hardness test machine (Emco Test DuraScan).

3.Results and discussion

Phase transformations of Cu-Al-Co alloy at different cooling conditions were measured by XRD measurements and shown in Fig. 1. The diffraction peaks were indexed according to the data given in the literature [12,13]. It was detected that as-cast and all other heat treated samples, except for furnace cooled one, possessed β'_1 martensite phase. The furnace cooled sample, however, had both β'_1 and α phases. It is considered that β phase could transform into both Cu-rich α fcc phase and Al rich β'_1-Cu_9Al_4 phase [12,14]. XRD diffraction analysis showed that this was also α phase.
Cooling rate yields a significant change in the microstructure of alloys. These morphological changes can be seen in Fig. 2. In Fig. 2 a, optical microscope images for as-cast samples are given. Here, grains can be seen clearly. The reason of the dark and bright appearance of the grains is the different orientation of the martensite plates. V-type and and needle-like martensites were dominant in the microstructure. Martensite plates aligned in parallel to each other.

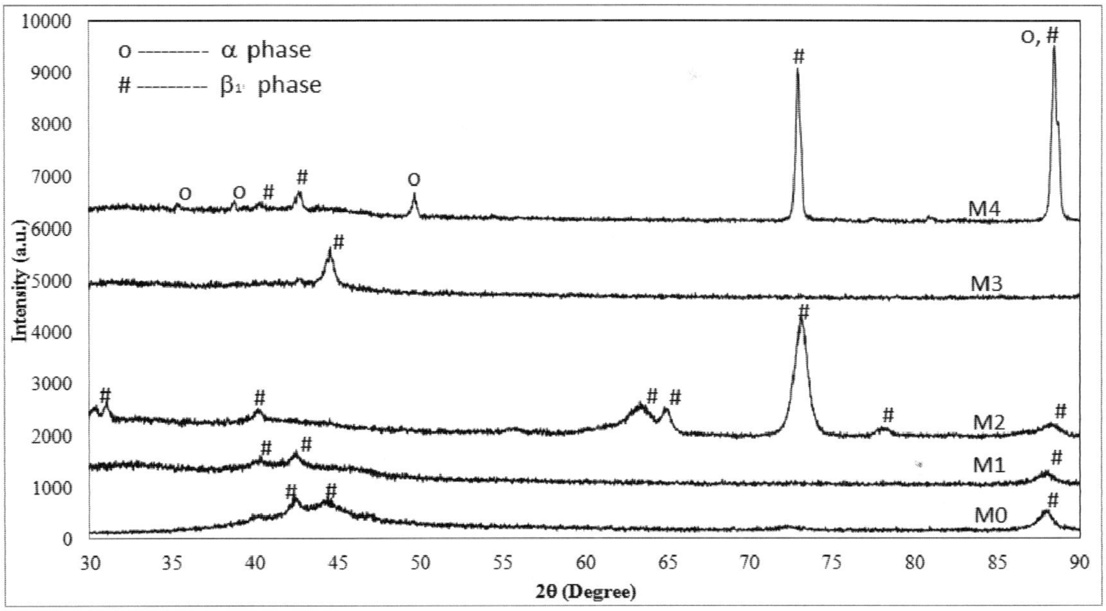

Figure 1. X ray difraktograms of Cu-Al-Co alloy at different cooling conditions.

The microstructure of the Cu-Al-Co alloy sample quenched in water with ice-salt mixture was similar to that of as-cast one (Fig 2 b.). Similar to as-cast sample, V-type, needle-like and plate martensites oriented in parallel to each other were observed in the microstructure. As seen from Fig. 2 c-d, there are similarities between the microstructure of the water quenched sample, the sample quenched into boiling water and the sample quenched into water with ice-salt mixture. Regularly aligned diamond shaped martensite plates can be observed in the structure. Microstructure of the furnace cooled sample obtained by the optical microscope, see Fig. e, showed that the microstructure had austenitic α phase that was also detected by XRD measurements. The dominant phase in the microstructure was α phase, while there was also minor amount of martensite (β'_1) in it. The variation in the transformation temperature of the Cu-Al-Co shape memory alloys, an indicator of the shape memory effect, subjected to different cooling conditions can be seen in Fig. 3. Meanwhile, transformation temperature values are summarized in the Table 1. According to the Table 1, there were significant differences between martensite start temperature of the Cu-Al-Co alloys, cooled at different conditions, while the austenite start temperature of these alloys did not change much.

The martensite start temperature was quite high for the furnace cooled sample. If the heat flow curve is examined for the same sample, it can be stated that DSC curve is broaden. It was considered that this broadening was caused by multiple phase transformation. As it can be revealed form the XRD plots, austenitic α phase and martensitic β'_1 phase appears simultaneously. This multiphase structure has been also detected in the furnace cooled NiTi alloy by Zhang and coworkers [10]. If the average enthalpy and entropy values are examined, the most striking difference can be observed for the furnace cooled sample. According to DSC measurements, transformation temperature of the alloy was significantly affect from the cooling rate and the presence of α phase in the furnace cooled sample reduced the transformation temperature

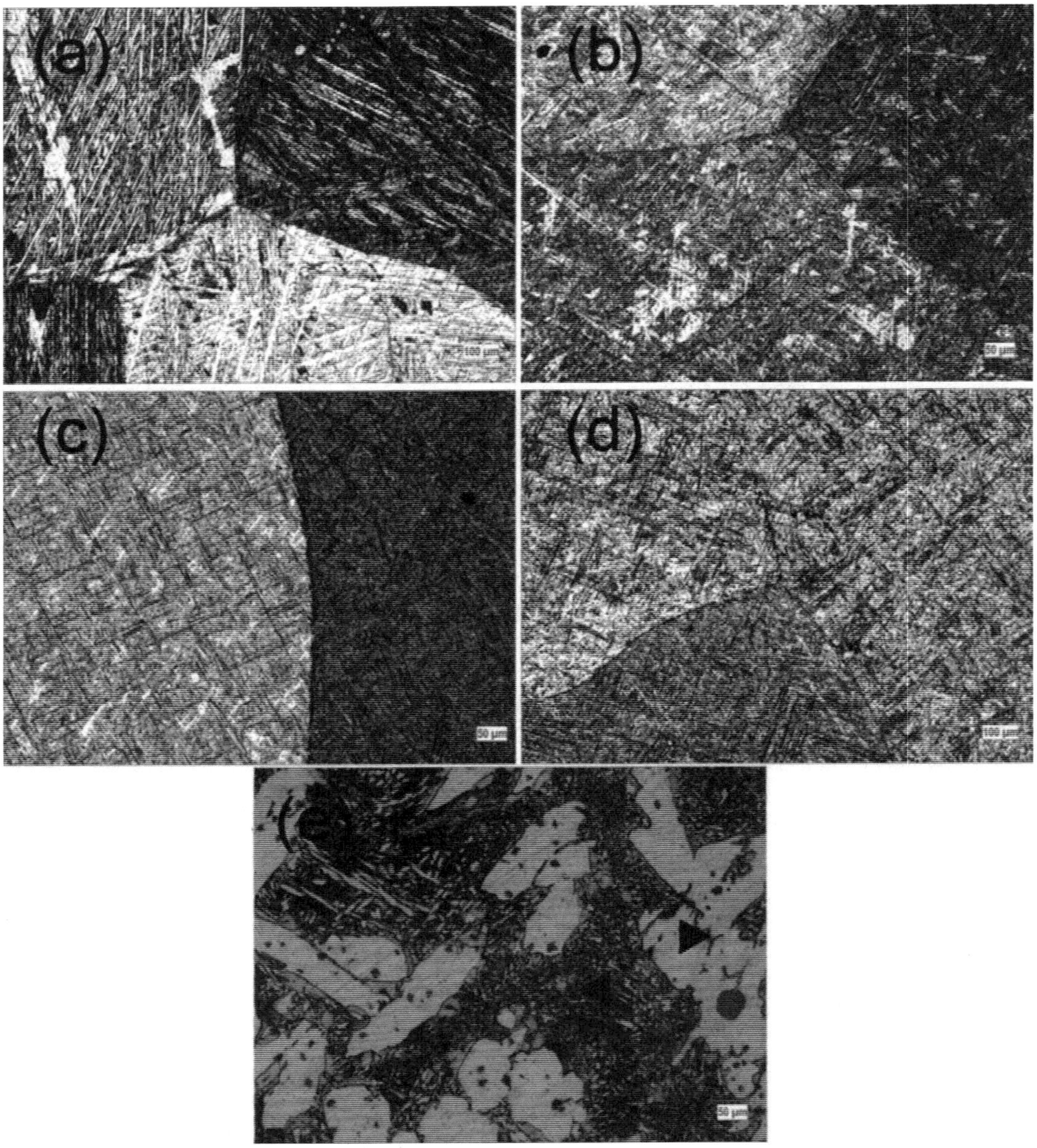

Figure 2. Optical micrographs of Cu-Al-Co alloy at different cooling conditions (◄ is β'_1 phase, ► is α phase) a) M0, b) M1, c) M2, d) M3 and e) M4

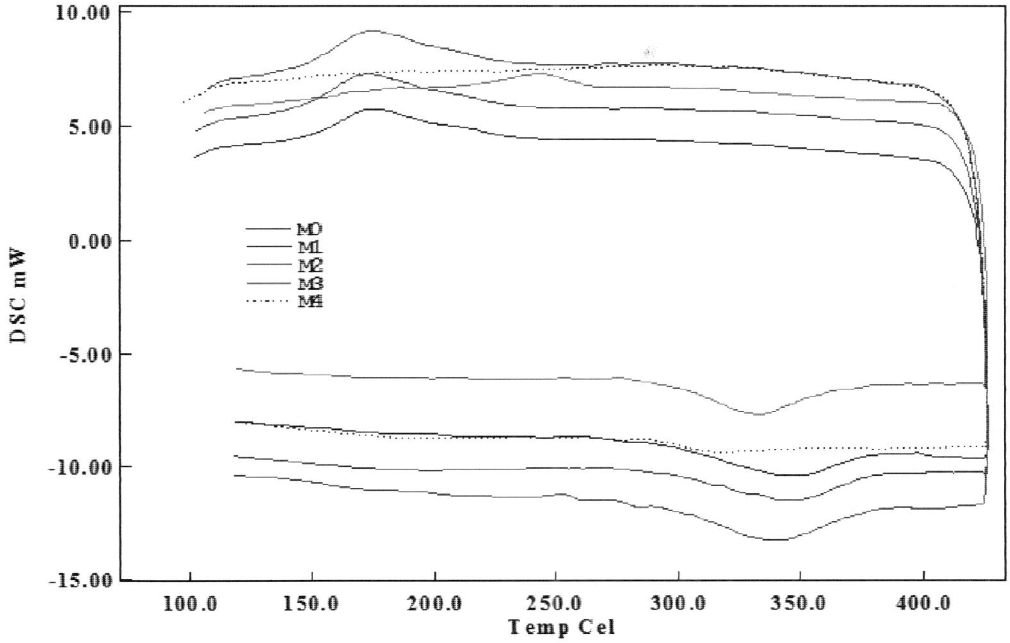

Figure 3. DSC curves of Cu-Al-Co alloy at different cooling conditions

Table 1. Phase transformation temperatures and micro hardness values of examined Alloys

Sample Code	A_s (°C)	A_p (°C)	A_f (°C)	M_s (°C)	M_p (°C)	M_f (°C)	ΔH_{ave} (J/g)	ΔS (J/g°C)	Average hardness (HV)
M0	295.3	333.1	357.4	264.9	243.1	217.3	2.76	0.010	263.3
M1	296.4	346.7	368.2	217.2	174.6	151.2	4.13	0.016	396
M2	294.7	346.3	376.7	226.9	173.1	148.5	4.28	0.016	309
M3	289.2	337.9	368.4	213.9	173.5	151.5	4.85	0.019	312
M4	286.5	317.9	343.2	350.1	292.8	239.7	1.32	0.004	257.3

Micro-hardness measurements are important as they give information about the brittleness of the shape memory alloys. Increasing hardness value could be considered as the reduction in the brittleness. Vickers hardness measurement scale is an important tool to understand the relation between microstructure and micro-hardness. The Vickers hardness value can be found by using the following equation, which includes the experimental data from micro-hardness measurement:

$$HV = \frac{2P\sin\emptyset/2}{d^2} \tag{1}$$

Here, $\emptyset$ is the apex angle of the indenter, P is the applied load and d is the average length of the diagonals [15].

For the same heat treatment temperature and duration, the variation of cooling conditions resulted in significant changes in the Vickers hardness value of the Cu-Al-Co alloy (Table 1). The lowest and highest hardness values were measured for the furnace cooled sample and the sample quenched to water with salt-ice mixture, respectively. Vickers hardness values of the samples quenched in a single step (directly into water) and quenched in two steps (first into boiling water for 3 minutes and then into water) were nearly equal but lower than hardness value of the sample quenched into water with ice-salt mixture. As a result of this, it can be revealed that the hardness value of the Cu-Al-Co alloy decreased with reducing cooling rate. Similar results have been also obtained for AISI 1020, AISI 1040 and AISI 1060 steels. In these studies, steels with four different carbon content has been furnace cooled, quenched in air or water after which the hardness values of the samples have been measured. It has been detected that the hardness values of the samples decreased with reducing cooling rate. This can be attributed to formation of the primary martensite phase upon rapid cooling [15-16].

4. Conclusions

It was observed that cooling conditions have significant effects on the transformation temperature, crystal structure, micro-hardness and microstructure of Cu-based shape memory alloys. It was detected that transformation temperature changed with cooling rate. The experiments showed that, with the application of furnace cooling, which was equivalent to the slowest cooling rate, cooling curve broadened. Meanwhile, the crystal analysis of this slowly cooled sample demonstrated the presence of α phase together with main martensite phase making the transformation more difficult to occur. Additionally, the presence of α phase was also verified by optical microscope analysis. It was measured that the micro-hardness value reduced with α phase.

Acknowledgement

This work has been supported by TUBITAK Research-Project No: 113F234

References

[1] Ma Y Q *et al* 2003 *Acta Metall Sin* **16-6** 445-448.
[2] Silva RAG *et al* 2013 *Thermochim Acta* **554** 71-75.
[3] Bublei I P *et al* 2008 *Met Sci Heat Treat* **11-12** 606-608.
[4] Bublei I P and Koval Y N 2006 *The Physics of Metals and Metallography* **101-4** 425-428.
[5] Dagdelen F *et al* 2003 *Mater Lett* **57** 1079-1085.
[6] Eskil M and Kayali N 2006 *Mater Lett* **60** 630-634.
[7] Ma L W *et al* 2012 *J Mater Eng Perform* **21-12** 2675-2679.
[8] Dunne D *et al* 2004 *Mater Sci Eng* **378** 257-262.
[9] Li Z *et al* 2008 *J Alloys Compd* **452** 307-311.
[10] Zhang Y Q *et al* 2012 *T Nonferr Metal Soc* **22** 2685-2690.
[11] Chandran P and Sampath V 2011 *Conference of Chemeca* 1-10.
[12] Aldırmaz E *et al* 2013 *Acta Physc Pol A* **124** 87-89.
[13] Celik H *et al* 2012 *Gazi University Journal of Science* **25-2** 337-342.
[14] Montecinos S *et al* 2009 *J Alloys Compd* **467** 278-283.
[15] Calık A. 2009 *Int J Phys Sci* **4-9** 514-518.
[16] Aydogdu Y *et al* 2015 9th *International Conference on Magnetic and Superconducting Material, 30thApril- 3rdMay, Antalya TURKEY*

Journal of Physics: Conference Series **667** (2016) 012011 doi:10.1088/1742-6596/667/1/012011

The change of transformation temperature on NiTi shape memory alloy by pressure and thermal ageing

M Kök[1], F Dağdelen[1], A Aydoğdu[2], Y Aydoğdu[2]

[1] Firat University, Faculty of Science, Department of Physics, Elazig, TURKEY.
[2] Gazi University, Faculty of Science, Department of Physics, Ankara, TURKEY.

E-mail: msoglu@firat.edu.tr

Abstract. We have investigated the effect of pressure and thermal ageing on transformation behavior of Ni-%45.16Ti shape memory alloy. 70-320 MPa pressure was applied and annealed at 500 °C for 2 hour to see the effect of applied pressure on transformation temperature. It was observed that the starting and final temperature of austenit increased, the martensite start temperature almost remain constant and final temperature decreased with the increase of pressure. The interval stress value of NiTi alloy was determined as 13.77 MPa/°C during austenite transformation. The transformation temperature of Ni-%45.16Ti alloys after thermal ageing was also investigated. No significant variation in these parameters was obtained. According to XRD measurements, there are two phases namely R (rhombohedral) and B19 (monoclinic) martensite were obtained. We did not see any change on the peaks after ageing.

1.Introduction

NiTi alloy with near-equiatomic composition is the most useful shape memory alloy (SMA) among many shape memory alloys. The application fields of these alloys are production of various medical devices such as orthopedic surgical implant, artificial prosthesis [1]. There are three different phases that have been observed throughout the martensitic transformation of NiTi: B2 (cubic), R (rhombohedral) and B19' (monoclinic). The high temperature austenitic B2 phase (CsCl) undergoes a displacive structure transformation to intermediate R phase about 293 K and transforms to the low temperature martensite B19' phase. It is not clear in which condition this unusual behavior occurs. This can be explained as another formation of martensitic behavior [2,3]. In summary, there are two series of phase transitions in NiTi:

$$B2 \rightarrow B19 \quad \text{and} \quad B2 \rightarrow R \rightarrow B19'.$$

The first series of transformations is observed in NiTi samples containing less than 50 %at Ni; the second occurs in samples with higher Ni concentration [4]. The shape memory behavior of NiTi alloys is influenced by their chemical composition, heat treatment (annealing and ageing) and mechanical deformation [5]. Effect of ageing treatment on the transformation behavior of NiTi alloy was studied by Y. Zheng and co-workers [6]. They chose various ageing temperature for 1.8 ks (30 min.) and investigated variation on martensitic transformation temperature and crystal structure. It was seen that the multistage phase appeared or disappeared by variable temperatures. Lin et al. were focused on ageing effect on V doped NiTi alloy. They aged this alloy at 400 °C for 1-10 h and examined martensitic transformation temperature and crystal structure changes. It was found that R phase appeared with residual ageing time [7]. K. Allafi and co-workers investigated the mechanism of multistage martensitic transformations in aged NiTi shape memory alloys. According to their study, annealed NiTi alloys at 850 °C for 15 minutes showed one step on both heating and cooling. When alloys exposed ageing at 400 and 500 °C for different times, multistage transformation was seen [8]. Chu et al. were focused on DSC study of aged porous Ni-rich NiTi shape memory alloy which fabricated by combustion synthesis. Their experimental study indicated that shape memory properties of porous NiTi alloy could be improved by optimizing the ageing technology [9]. The understanding of strain and phase transformation behavior under constant applied pressure is essential, because of improving recent application of NiTi shape memory alloys, including smart structures, intelligent controllers and memory device [10]. There is few study about effect of constant pressure on shape memory alloys. Kröger et al. studied pressure-induced martensitic transformation in Ni rich NiTi alloys [11]. They measured DSC pressure-free and pressure-assisted aging for 4 ks (almost an hour) at 550 °C. It was seen that martensitic transformation temperature of NiTi alloys increased after pressure. The aim of this study is to investigate the effect of pressure and thermal aging on the martensitic transformation of Ni-%45.16Ti shape memory alloys.

2. Experimental

NiTi shape memory alloy with nominal composition of Ni-%45.16Ti (atomic) was supplied by Nimesis technology. All samples were annealed for 2 hour at 500 °C and rapidly quenched in ice brined. DSC measurements were done in order to investigate the effect of pressure and thermal ageing on transformation temperature. These temperatures namely Martensite finish-M_f , Martensite start-M_s, Austenite start-A_s and Austenite finish-A_f of alloys were measured by Perkin Elmer Sapphire differential scanning calorimetry with 10 °C/min heating and cooling rate at the temperature range of 0 and 150 °C in Nitrogen atmosphere. X-ray analysis system (Bruker D8 Advance) using Cu K_α radiation was used to determine crystal structure of thermal aged alloys. The X-ray diffraction (XRD) measurements were carried out in the range of 20-80° at the rate of 2°.

3. Result and discussion

The DSC charts on heating and cooling of the NiTi shape memory alloy annealed at 500 °C were shown in Figure 1. According to Figure 1, two phase transformation occur during cooling. One of these phase is martensite other is rhombohedral (R) phase.

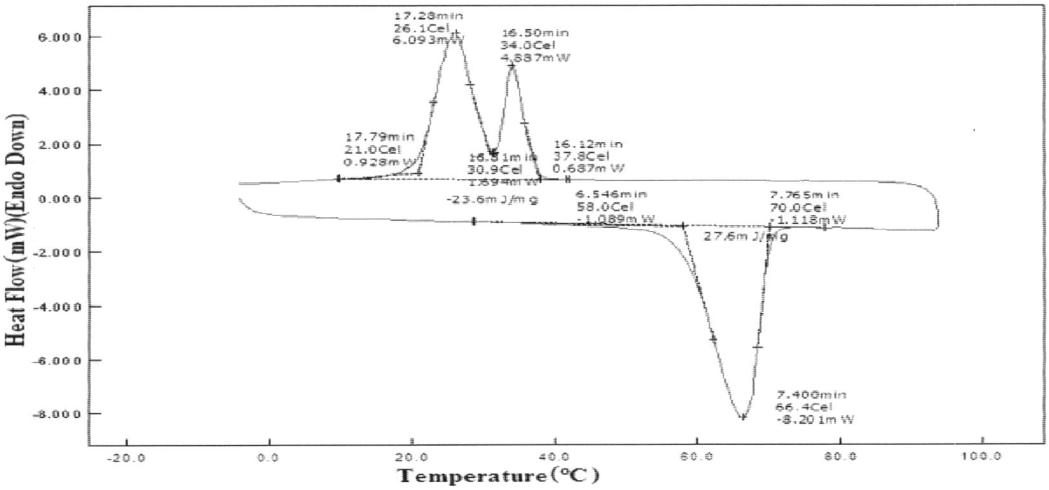

Figure 1. DSC thermogram of NiTi alloy annealed at 500 °C

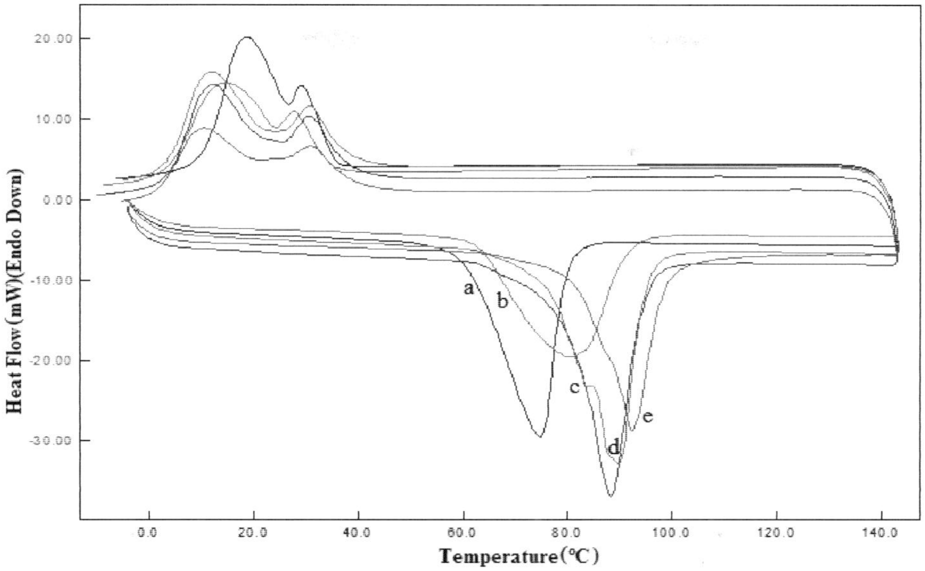

Figure 2. DSC thermograms of a) control NiTi alloy b) loaded NiTi alloy under 78 MPa pressure c) loaded NiTi alloy under 156 MPa pressure d) loaded NiTi alloy under 234 MPa pressure e) loaded NiTi alloy under 312 MPa pressure.

It is known fact that the phase transformation steps of NiTi alloy depends on heat treatment temperature: A↔R at annealing temperatures below 400 °C, A↔R↔M at temperatures 400-600 °C and A↔M at temperatures above 600 °C [12,13]. R phase transformation is related to formation of Ti_3Ni_4 precipitates [7].

3.1. Pressure ageing

Figure 2 show the DSC thermogram of unloaded and loaded NiTi shape memory alloy during heating and cooling. Transformation temperatures of the material are summarized in Table 1. Phase transformation temperatures (A_s, A_f, M_s, M_f), Temperature Transition Region (TTR) which is equal to A_f - M_f and the variation of thermodynamics parameter (enthalpy and entropy) are given together.

Table 1. Transformation temperatures and thermodynamic parameters list of pressure aged NiTi.

	No-Pressure	78 MPa	156 MPa	243 MPa	312 MPa
A_s (°C)	58.0	61.7	75.5	74.4	80.4
A_p (°C)	66.4	80.2	88.3	89.6	92.3
A_f(°C)	70.0	88.3	94.8	95.2	99.0
M_s(°C)	37.8	34.7	39	38.1	40.3
R_p (°C)	34.0	28.2	30.8	30.7	30.9
M_p (°C)	26.1	14.7	12	11.9	10
M_f (°C)	21.0	3.7	4.1	2.7	1.8
T_o (°C)	47.9	48.2	57.2	56.2	60.3
TTR	49.0	84.6	90.7	92.5	97.2
$\Delta H_{heating}(Jg^{-1})$	27.6	24.1	25.4	24.8	22.4
$\Delta H_{cooling}(Jg^{-1})$	-23.6	-19.2	-19.8	-19.4	-17.4
$\Delta S(Jg^{-1o}C^{-1})$	0.53	0.45	0.39	0.39	0.33

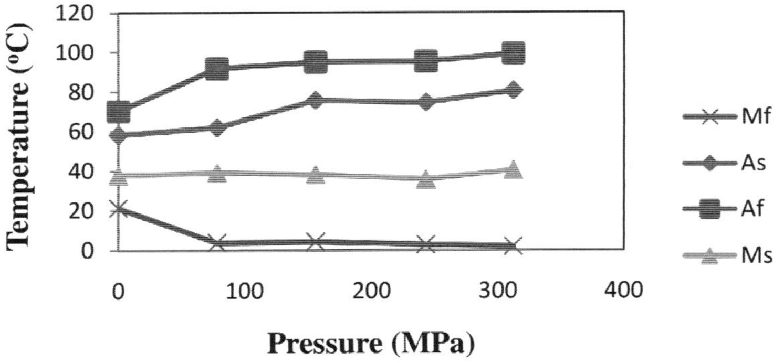

Figure 3. The change of transformation temperature with applied pressure

Transformation temperatures generally increase with the rising amount of pressure. Besides, decomposition occurred at transformation temperature. The electron distribution rises with the shrinkage of the atomic distance by the pressure; hence a significant change occurs at transformation temperature. Evaluation of critical temperatures is presented in the Figure 3. The austenite start and final temperature increase, the martensitic start temperature remains almost constant and martensitic final temperature decrease with the rising pressure. The reason of decreasing in martensite final temperature may be the increasing of Ni concentration in NiTi matrix or the growth of the Ti-rich phase by applied pressure [14].

It was seen by evaluating the thermodynamic parameters given in Table 1 that the average enthalpy values obtained for the alloys subjected to pressure ($\Delta H_{average}=(\Delta H_{heating}+\Delta H_{cooling})/2$) increases than that of not to subject pressure. Besides a reduction were determined in entropy variation with the pressure. These indicate that reduction in entropy variation can induce a reduction in unsteady structure of NiTi alloy. Clausius-Clapeyron equation for austenite transformation can be written as

$$\frac{d\sigma}{dT} = -\frac{\rho \Delta H}{\varepsilon T_o} \tag{1}$$

where ρ is density of NiTi alloys and equal 6500 kg/m^{-3}, ΔH is enthalpy change, $T_o= (A_s+M_s)/2$ is equilibrium temperature, σ is the pressure and ε is the strain associated with transformation. The strain values are proportional to transformation temperature and especially to T_o temperature at which martensite and austenite transformation are equal to Gibbs free energy according to Clausius-Clapeyron equation [15]. The T_o value increases with the pressure as seen in Table 1.The strain values also increase with the pressure. Clausius-Clapeyron equation can be used another form for the previous equation:

$$\frac{d\sigma}{dT} = -\frac{\rho \Delta S}{\varepsilon} \tag{2}$$

ΔS is entropy change and equal to $\Delta H/T_o$. DSC measurements give enthalpy change of NiTi alloys during heating or cooling [14]. The define Stress rate described at the left side of the equation for the stress induced transformation was determined as 13.77 MPa/°C during austenite transformation. For NiTi alloys the stress range is between 3 and 20 MPa °C^{-1} [16]. The results are in accordance with the literature.

Table 2. Transformation temperatures and thermodynamics parameters list of thermal aged NiTi Alloys.

Ageing Time at 200 °C	0h	1h	2h	3h	4h	5h	6h	7h
A_s (°C)	58.0	58.2	58.0	58.6	58.7	59.3	59.0	59.1
A_p (°C)	66.4	65.6	64.5	65.6	65.6	69.0	66.1	67.3
A_f(°C)	70.0	68.3	67.4	68.9	68.7	72.5	68.7	70.3
M_s(°C)	37.8	37.6	38.2	38.4	38.0	37.6	38.2	38.5
R_p (°C)	34.0	34.5	35.6	35.1	35.1	33.2	35.2	34.3
M_p (°C)	26.1	26.7	28.5	28.1	28.1	24.9	28.3	26.6
M_f (°C)	21.0	22.4	24.3	22.9	23.1	18.6	23.7	21.1
T_o(°C)	47.9	47.9	48.1	48.5	48.5	48.4	48.6	48.8
$\Delta H_{heating}$ (Jg^{-1})	27.6	26.6	27.0	26.6	27.0	24.9	26.8	25.4
$\Delta H_{cooling}$(Jg^{-1})	-23.6	-22.6	-22.9	-22.7	-23.2	-22.3	-22.8	-22.2
ΔS (Jg^{-1}°C^{-1})	0.53	0.51	0.52	0.51	0.52	0.49	0.51	0.49

3.2. Thermal ageing

The DSC measurements for investigating the effect of thermal ageing depending on time are summarized in Table 2. The austenite start temperatures A_s of Ni-Ti alloy are almost same with the changed ageing time but there is an increase in austenite final temperatures A_f of Ni-Ti alloy at 5 hour ageing (Figure 4). This increase in thermal transformation temperature can be explained as the instantaneous change in entropy of the aged alloy or this feature is believed to be related to the variation of Ni/Ti ratio in the NiTi matrix [7]. If, Ni/Ti ratio decreases in matrix, the transformation temperatures increase. So, transformation temperatures are very sensitive to the composition of the NiTi alloy and increase with decrease of Ni concentration [7, 17].

The martensitic start temperature M_s of alloy remains almost constant as well austenite transformation. The opposite of A_f, the martensitic final temperature M_f of alloy fall down at 5h ageing time (Figure 4). As seen Table 2, presence of R phase, there was difference at austenite and martensitic final temperature and enthalpy change, for the 5h aged time. The biggest A_s and enthalpy change were seen in this ageing time. The calculated ΔS and T_o values were given in Table 2. ΔS and T_o values were almost constant with rising ageing time.

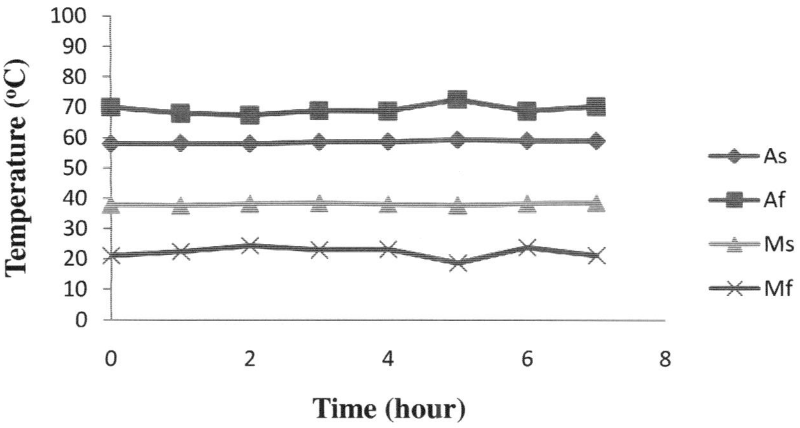

Figure 4. The change of transformation temperature with ageing time

Fig. 3 and Fig. 4 represent the effect of pressure and thermal ageing on transformation temperature respectively. It is seen from the figures that an observable change in transformation temperature occurs by the applied pressure. There is no significant variation at 200 °C temperature and 0-7 hour interval. X-ray patterns of thermal aged NiTi alloys at room temperature were given in Figure 5. The X-ray pattern of unaged alloy demonstrates the main peak occurred at about 42.5° (-101) with minor peaks at about 41° (-111), 42° (110), 43.5° (002) and 44.5° (111). These are the characteristic peaks and in agreement with the peaks observed by Czeppe et al. [18]. It is seen that there are two phase in crystal structure of unaged NiTi shape memory alloy. The main peak (-101) and at 42° (110) belong to rhombohedral (R) phase, other peaks belong to monoclinic (B19) phase. All XRD pattern exhibited similar characteristics according to investigation of the changes in peak characteristics of XRD patterns with ageing duration [19]. However, for one hour ageing time, some peaks such as 43.5° (002)B19 and 44.5° (111)B19 disappeared.

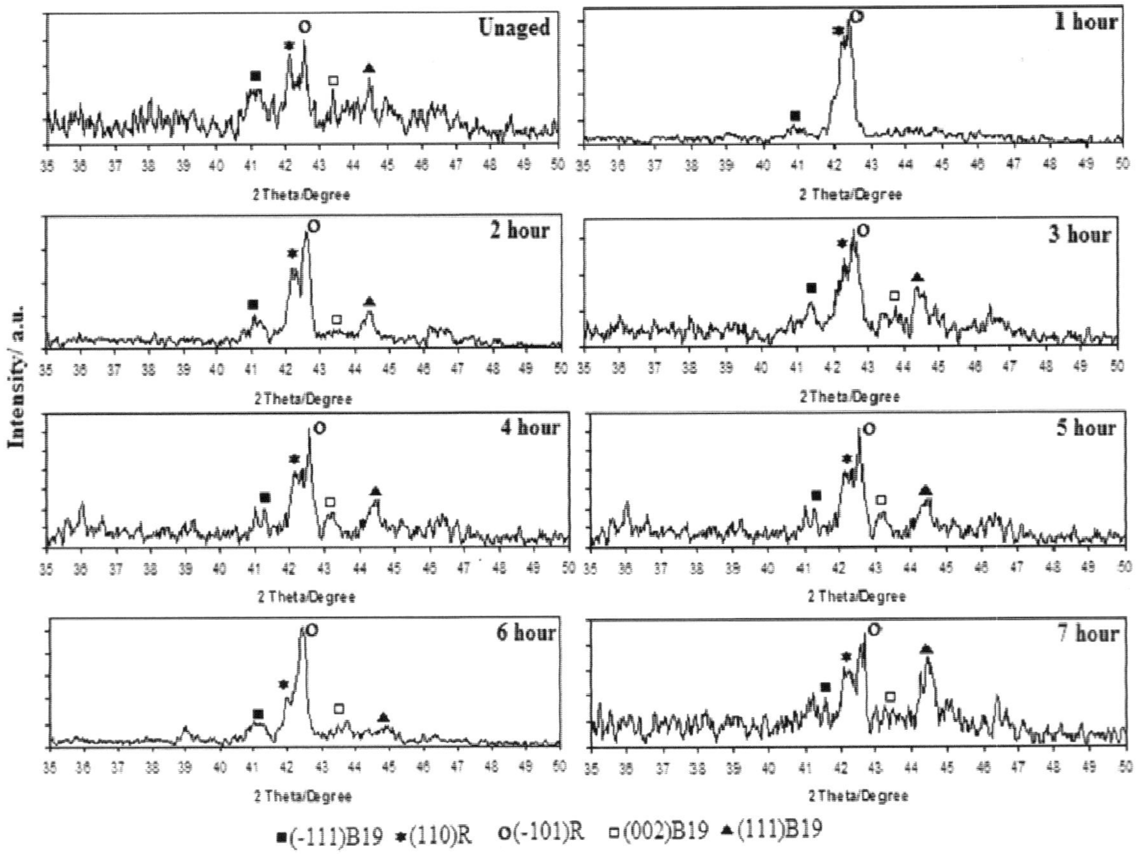

Figure 5. XRD diffractograms of thermal aged NiTi alloy

4. Conclusion

The results obtained from this study can be outlined as follows:

- It was determined that while the transformation temperature increased the entropy change decreased by the pressure ageing. The stress interval value was determined as 13.77 MPa/°C during the austenite transformation temperature.
- When ageing time effect on transformation examined at 500°C annealing temperature, the stable position was seen for 200°C ageing temperature. The X-ray analyses showed that there was two phase as R (rhombohedral) and B19 (monoclinic) martensite. There was not important change in characteristics peaks by ageing process.

In conclusion it was observed by comparing the pressure and thermal ageing that while there is a significant variation in transformation temperature with the pressure, there is not an observable change with the chosen ageing temperature.

Acknowledgements

This work is supported by TUBITAK under Project No: 113F234 and Firat University Research-Project unit under Project No: FF. 12.08.

References

[1] Sadrnezhaad K *et al* 1997 *Mater. Manuf. Process.* **12-1** 107-115
[2] Zeng Z Y *et al* 2010 *Pysica B* **405** 3665-3672
[3] Wang Z G *et al* 2004 *Mater. Lett.* **58** 3141-3144
[4] Egorushkin V E and Kulkova S E 1982 *J. Phys. F Met Phys* **12** 2823-2888
[5] Sadiq H *et al* 2010 *Smart Mater. Struct.* **19**1-7
[6] Zheng Y *et al* 2008 *Acta Mater.* **56** 736-745
[7] Lin H C *et al* 2008 *J. Alloy Compd.* **449** 119-124
[8] Khalil A J *et al* 2002 *Acta Mater.* **50** 793-803
[9] Chu L *et al* 2005 *Mater. Lett* **.59** 404-407
[10] Yan X and Humbeeck J V 2011 *Smart Materials Research* doi:10.1155/2011/160927
[11] Kröger A *et al* 2008 *Mater. Sci. Eng. A* **481-482** 452-456.
[12] Choi J Y and Nemat-Nasser S 2006 *Mater. Sci. Eng. A* **432** 100-107
[13] Sadrnezhaad S K and Selahi A R 2004 *Mater. Manuf. Process.* **19-4** 475-486
[14] Yamamoto T *et al* 2007 *Mater. Trans.* **48-3** 439-444
[15] Daroczi L *et al* 2000 *Mater.* **43** 691-697
[16] Pelton A R *et al* 2000 *Min. Invas. Ther. And Allied Technol.* **9-1** 107-118
[17] Chang C Y *et al* 2001 *Metall Metal Trans. A* **32** 1629-1634
[18] Czeppe T *et al* G 2009 *Vacuum* **83** S214-S219
[19] Kök M *et al* 2015 9th *International Conference on Magnetic and Superconducting Material, 30thApril- 3rd May, Antalya TURKEY*